Explorations Using the TI-82 & TI-83

Beverly K. Michael
University of Pittsburgh

to accompany

Explorations in College Algebra

Linda Almgren Kime
Judith Clark
University of Massachusetts, Boston

JOHN WILEY & SONS, INC.
New York • Chichester • Weinheim • Brisbane • Singapore • Toronto

PREFACE

To the student

- Each chapter in this *Graphing Calculator Workbook* is designed to correspond with each chapter in *Explorations in College Algebra.* The exception is Chapter 0, which is to get you started.
- There is no need to learn everything at once. Calculator techniques are introduced as you need them in each chapter.
- It is assumed that you are using either a TI-82 or TI-83 Graphing Calculator and that you are working right along with the author.
- Since the text exercises assumes that you have either a computer or graphing calculator available, you will want to keep the *Graphing Calculator Workbook* available as a reference when you are doing the exercises.
- You should work through each chapter in the *Graphing Calculator Workbook* before you attempt to do the text exercises.
- Your instructor may assign the *Graphing Calculator Workbook* chapters for homework, for an inclass activity or for an inclass demonstration. Whatever technique is used, follow along with your own calculator. One will learn techniques best by doing.
- The "rule of three" is followed throughout the *Graphing Calculator Workbook.* Mathematical ideas are introduced graphically, numerically and algebraically. For complete understanding you should be able to solve a mathematical problem in a variety of ways.
- A graphing calculator is a tool. It is only as good as its operator. You should practice using this tool for the right job. One of the best ways to use the graphing calculator is as a verification tool to confirm your answers. Another is as a tool for performing mathematical experiments by creating "what if" situations.
- Learning anything should be fun. Enjoy yourself. Remember confusion is the first step to learning, if you are not confused then you are at a practice stage and practice makes perfect.

Acknowledgments
Thank you to:
- Linda and Judy for being inspired to write their book.
- Dan, Megan and Bridget for their patience and love.
- Frank and Bert for sharing their enthusiasm for graphing technology.

Beverly K. Michael
bkm+@pitt.edu

Contents

Chapter 0
Getting Started on the TI-82 or TI-83[1]

0.1 Turn the Calculator ON / OFF
Locating the keys.

Turn your calculator on by using the |ON| key, located in the lower left- hand corner of the calculator. To turn the calculator off press |2nd| |OFF| : located above the |ON| key.

To locate the correct keys think of your calculator as being divided into three sections:

1. The bottom six row of keys are your mathematical calculation and function keys.
2. Rows 7 - 9 are the menu and editing keys.
3. The very top row (under the screen) is where your graphing keys are located.

0.2 Adjusting the Screen Contrast

Depending on the room lighting you may want to adjust the screen contrast.

1. To darken the screen:

Press and release the |2nd| key, then press. and hold the up arrow |Δ| key.

2. To lighten the screen:

Press and release the |2nd| key, then press and hold down arrow |∇| key.

As the display contrast changes, a number appears in the upper right corner of the screen between 0 (lightest) and 9 (darkest).

If you adjust the setting to 0, the display may become completely blank. If this happens, increase the contrast and the display will reappear.   When contrast needs to be set at 8 or 9 all the time, it is probably time to change the batteries.

0.3 MODE Default Settings

The calculator should be set to the default mode settings. Press |MODE| to see the settings.

Set your calculator to the settings in Figure 0.1 or 0.2 using your arrow keys and pressing |ENTER| to activate your choice.

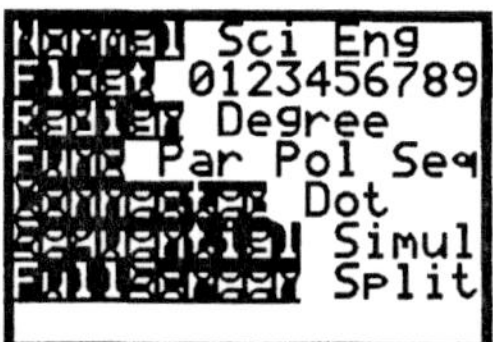

Figure 0.1
The TI-82 default mode screen.

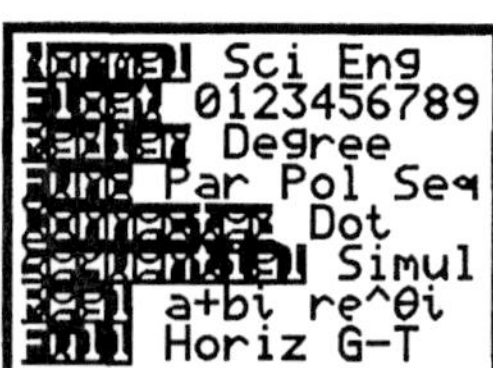

Figure 0.2
The TI-83 default mode screen.

Note: If your calculator is not new you may want to RESET MEMORY. This will completely erase all data and programs and reset the calculator to the default mode. Use this cautiously.
Press |2nd| |MEM| (above +),select **3** then select **[2:Reset]**.

---

[1] The key stroking and menus for the TI-82 and TI-83 are nearly the same. Where they differ, screen images are presented side by side.

0.4 The Home Screen

The Home Screen is your calculation and execution of instruction screen. To return to the Home Screen from any other screen, press [2nd] [QUIT] . The Home Screen is the primary screen of the TI-82 or TI-83. If there is something displayed on the Home Screen, press the [CLEAR] key.

0.5 Calculating

The bottom six rows of keys on the graphing calculator behave like those on any scientific calculator, except that your entry is seen on an eight line computer screen. When you want the calculator to perform any calculation or instruction, press [ENTER] .

Example 1
From the Home screen, do the following:

1. Type 12 [X] 2 then press [ENTER] ; 24 is now displayed and *stored* as the answer. See Figure 0.3.

2. Press [2nd] [ANS] and [ENTER] ; 24 is again displayed.

3. Press the multiplication key [X] , then 2 and then [ENTER] . Pressing any operation key, +, -, X, ÷, x^2, x^{-1} etc. , assumes that you want to operate on the stored answer (see Fig. 0.3).

0.6   Iteration, Recalling a Process

Notice how [ANS] is also used.

Repeatedly press [ENTER] . Your screen should look like the bottom of Figures 0.3 and 0.4.

This process is called <u>iteration</u> (repeating some process over and over again). The last operation (multiplying by 2) is repeated on the new answer.

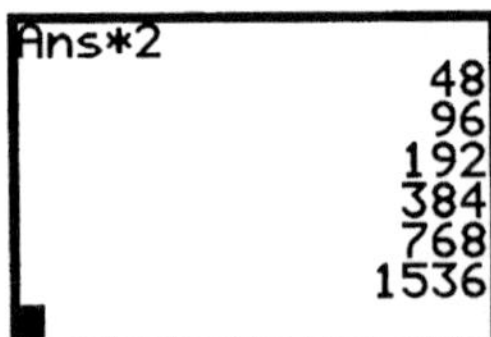

```
12*2
            24
Ans
            24
Ans*2
            48
            96
```

Figure 0.3

The asterisk , * , is used for multiplication in place of the "times" sign to avoid confusion with the letter x**.**

```
Ans*2
            48
            96
           192
           384
           768
          1536
```

Figure 0.4

Example 2

Interest compounded at 5% annually on an initial investment of $1000 can be represented by 1000*1.05, or A = P(1 + R) for the first year .

[Amount = original investment(1 + rate).] Use iteration to determine the number of years for the amount of accumulated investment to be greater than $1300.

Press CLEAR to clear the Home Screen.

Type 1000 followed by ENTER .

The number 1000 is now stored in memory.

Press X 1.05 ENTER . The number 1050 will now be displayed. By repeatedly pressing ENTER , you can see the growth of your initial $1000 investment year by year and determine that 6 iterations (years) are necessary for you to exceed $1300. See Figures 0.5 and 0.6.

0.7   Converting Decimals and Fractions

The TI-82 or TI-83 can be used to convert decimals and fractions.

Press 1 ÷ 4 ENTER (see Fig. 0.7).

The decimal answer for this expression, 0.25, is displayed. Press MATH . You are in the MATH menu . Menus give a list of additional command options (see Fig. 0. 8 or 0.9). Press 1 or ENTER to select the highlighted option. This option [1: Frac] will change the decimal answer back into a fraction.

Note: When the denominator of a fraction has more than four digits the answer is displayed as a decimal and will not return to a fraction.

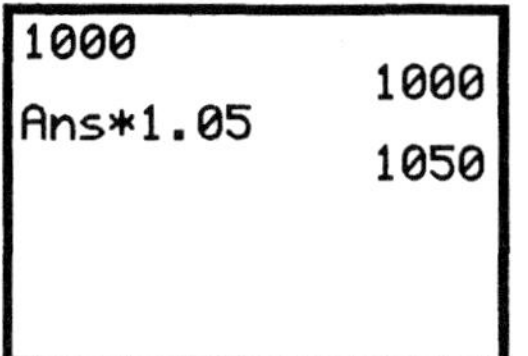

Figure 0.5

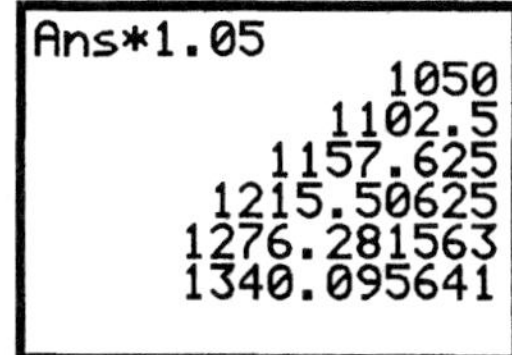

Figure 0.6

Between year 5 and 6 the amount is > 1300.

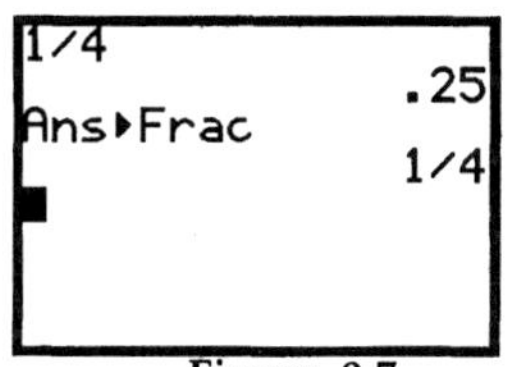

Figure 0.7

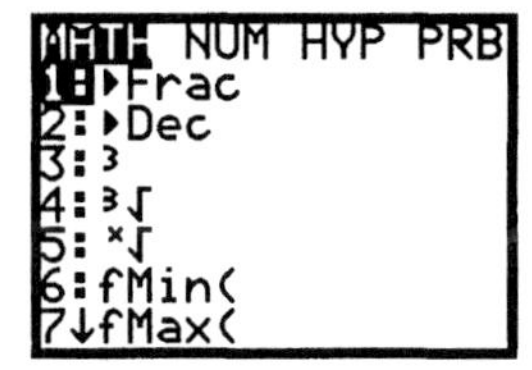

Figure 0.8

The TI-82 MATH menu

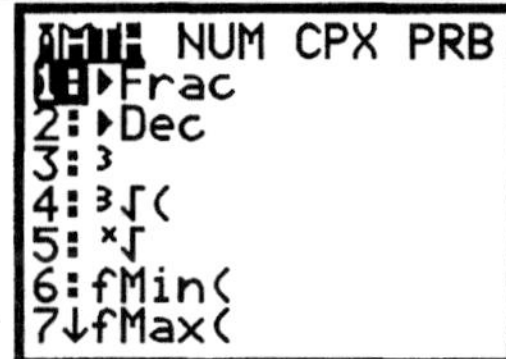

Figure 0.9

The TI-83 MATH menu

0.8 Selecting Items from a Menu.

You can select an item from a menu by typing the number or by moving to that menu option with the down arrow key $\boxed{\nabla}$.

You press $\boxed{\text{ENTER}}$ to select your menu option. Press $\boxed{\text{MATH}}$ $\boxed{\nabla}$ select [2: Dec] .

Press $\boxed{\text{ENTER}}$, this changes the fraction back to a decimal (see Fig. 0.10).

Example 3
Type in the following fraction problems, then use the MATH menu to change the answers back to fractional form.

1. $\dfrac{1}{2}+\dfrac{1}{3}$

    Press $(1\boxed{\div}\ 2)\ \boxed{+}\ (1\boxed{\div}\ 3)\ \boxed{\text{ENTER}}$.

    Press $\boxed{\text{MATH}}\ \boxed{1}\ \boxed{\text{ENTER}}$.

2. $3\dfrac{5}{9}+5\dfrac{3}{7}$

Press $\boxed{(}\ 3\boxed{+}\ 5\boxed{\div}\ 9\boxed{)}\ \boxed{+}\ \boxed{(}\ 5\boxed{+}\ 3\boxed{\div}\ 7\boxed{)}$

$\boxed{\text{ENTER}}$. Press $\boxed{\text{MATH}}\ \boxed{1}\ \boxed{\text{ENTER}}$

(see Fig. 0.11).

The answer to Example 3 part 2 is 566/63.

Press $\boxed{\text{CLEAR}}$.

0.9   Raising a Number to a Power

The calculator can be used to raise a number (called the base) to a power by using the exponent key, $\boxed{\wedge}$.

For 3^2 press $3\ \boxed{\wedge}\ 2\ \boxed{\text{ENTER}}$ or use a short cut press $3\ \boxed{x^2}\ \boxed{\text{ENTER}}$. This last method pastes the exponent to the right of 3 (see Fig. 0.12).

Example 4
Type the expression:

$$3^4\ \text{x}\ 2^5 \div 6^2$$

Press $\boxed{\text{ENTER}}$ for the result (see Fig. 0 12).

0.10 Order of Operations

The TI-82 or TI-83 uses an algebraic order of operations: inside parentheses first, powers next, then multiply or divide from left to right and lastly add or subtract from left to right.

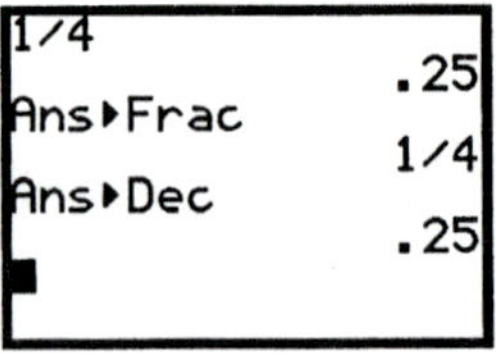

Figure 0.10

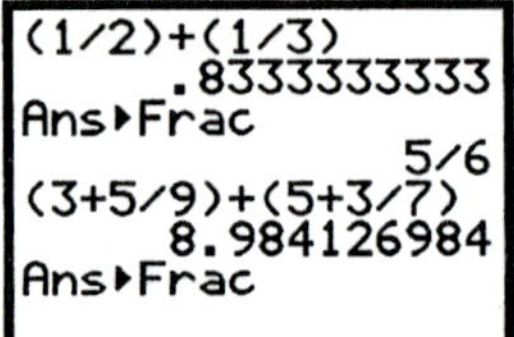

Figure 0.11

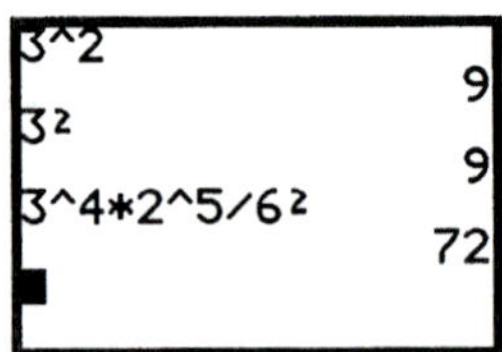

Figure 0.12

Example 5

1. **Enter: $1 + 2(4 - 2)^2 + 6 \div 2$**
(see Figure 13).
The order of operations are performed algebraically in the following steps:

$1 + 2(4 - 2)^2 + 6 \div 2 =$	
$1 + 2(\mathbf{2})^2 + 6 \div 2 =$	inside parentheses
$1 + 2(\mathbf{4}) + 6 \div 2 =$	raise to power
$1 + \mathbf{8} + 6 \div 2 =$	multiply
$1 + 8 + \mathbf{3} =$	divide
$\mathbf{9} + 3 =$	add
$\mathbf{12}$	add

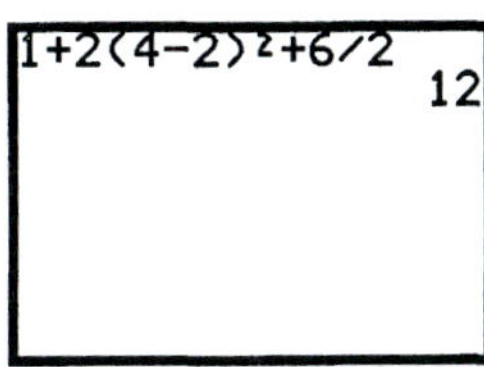

Figure 0.13

2. **Enter: "One hundred fifths times two."**
See Figure 0.14 for two methods.

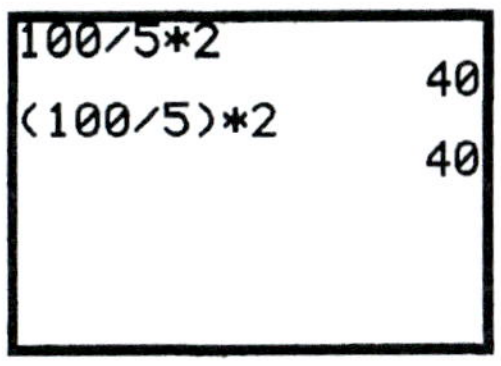

Figure 0.14

<u>Troubleshooting</u>: For the TI-82 users, compare the difference in order of operations in Figures 0.14 and 0.15. Parenthesis in the denominator of a fraction are interpreted as a grouping . For the TI-83 parentheses are interpreted the same as the multiplication sign (see Fig. 0.16). To avoid confusion always enclose fractions in parentheses.

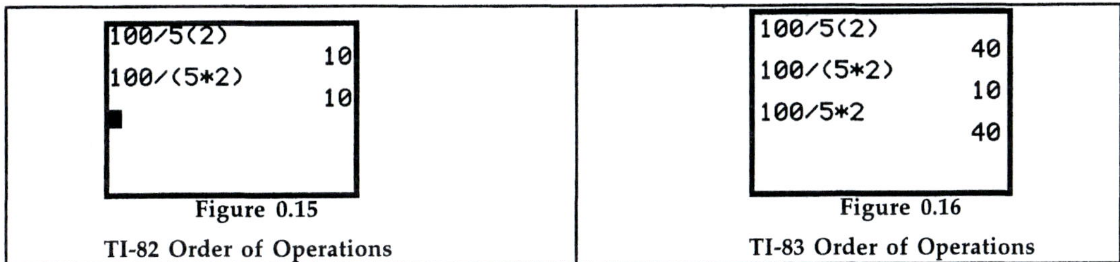

Figure 0.15

TI-82 Order of Operations

Figure 0.16

TI-83 Order of Operations

3. **Enter "Sixteen raised to the one half power."**
This is the same as "the square root of 16."

<u>Note</u>: 16^1/2 is ***not*** √16; fractional exponents must always be enclosed in parentheses (see Fig. 0.17).

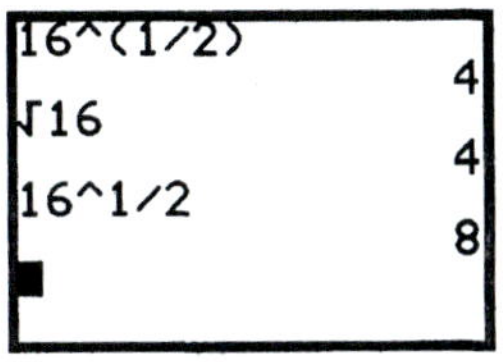

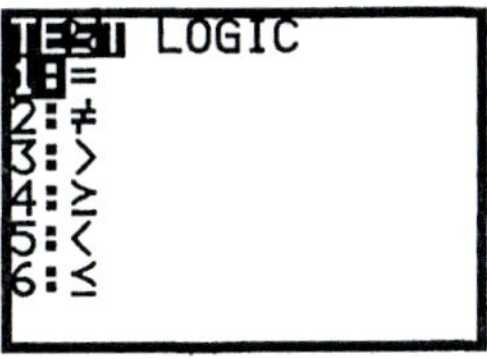

Figure 0.17

0.11 Truth Tests

The graphing calculator can be used to determine whether an expression is true or false.

To use this feature, you must use the [2nd] [TEST] menu. Figure 0.18 shows the TEST menu. This is where the equal and inequality symbols are located.

Figure 0.18

Example 6

1. Is $3 < 7$ **true or false?**

Press 3 | 2nd | | TEST | | V | .

Select [5:<.] ,press 7 | ENTER | (see Fig. 0.19).

<table><tr><td style="border:2px solid black; padding:6px;"><u>Note</u>: When performing a TEST, remember that 1 means TRUE and 0 means FALSE.</td></tr></table>

2. **Is** $3(4 + 5) = (3 \times 4) + 5$ **true or false?**

This is a false statement, thus the answer is zero (see Fig. 0.19) because:
$$3(4 + 5) = (3 \times 4) + (3 \times 5)$$

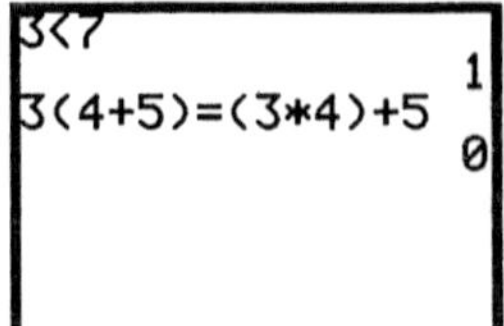

Figure 0.19

0.12 Deep Recall and Editing

Press | CLEAR | . To recover your last entry press | 2nd | | ENTRY | . To evaluate press | ENTER | . To edit an expression, use the left and right arrows to position the cursor for editing and press delete | DEL | or insert | 2nd | | INS | .

Example 7

Change the expression in Example 6 part 2 to : $3(4 + 5) = (3 \times 4) + (3 \times 5)$.

First recall the expression, | 2nd | | ENTRY | .

Use | ◁ | to place the cursor on the 5; press | 2nd | | INS | type | (| 3 | X | . Then | ▷ | to place the parentheses after the 5, | ENTER | (see Fig. 0.20). Now the expression is evaluated as true   (i.e. the number 1).

<table><tr><td style="border:1px solid black; padding:6px;">3(4+5)=(3*4)+5
0
3(4+5)=(3*4)+(3*
5)
1</td></tr></table>

Figure 0.20

<table><tr><td style="border:2px solid black; padding:6px;"><u>Note</u>: Also try pressing | 2nd | | ENTRY | several times and you will see some of the old expressions you typed. This is called <u>deep recall</u> and it is used to retrieve expressions that have been typed many steps earlier.</td></tr></table>

0.13 Storing Values to Variables

Recall Example 2 where we were finding the amount of money accumulated after one year (A) using the formula $A = P(1 + R)^x$, where the principle P = \$1000 and the rate R = 5%, and x=1. The calculator allows you to store values to alphabetical letters A through Z. You access the letters by first pressing the $\boxed{\text{ALPHA}}$ key and you store number values to letters by using the store $\boxed{\text{STO}}$ key.

> Note: Alphabetical letters are located to the above right of keys and are color coded to match the $\boxed{\text{ALPHA}}$ key.

Example 8
Find A if P = 1000 , R = .05 and x=1 using
$$A = P(1 + R)^1 = P(1 + R).$$

1. To store 1000 to P , press 1000 $\boxed{\text{STO}}$
   $\boxed{\text{ALPHA}}$ $\boxed{\text{P}}$ $\boxed{\text{ENTER}}$.

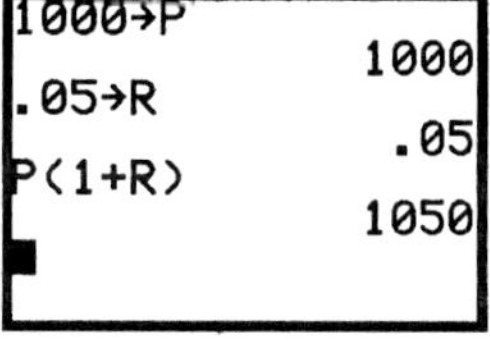

Figure 0.21

2. To store .05 to R , press .05 $\boxed{\text{STO}}$
   $\boxed{\text{ALPHA}}$ $\boxed{\text{R}}$ $\boxed{\text{ENTER}}$ (see Fig. 0. 21).
3. Type the expression $\mathbf{P(1 + R)}$;
   remember to press $\boxed{\text{ALPHA}}$ before
   typing the letter, and press $\boxed{\text{ENTER}}$.

The expression has been evaluated using the stored values to P and R. These values will remain the same until you store a new value to R and P (see Fig. 0. 21).

> **Trouble Shooting**: If your calculator is new or if the memory has been cleared, the initial stored value to all letters is zero.

0.13.1 A special note about x and y

Since the variables x and y are used in plotting graphs, their values are constantly updated when you TRACE on a graph.

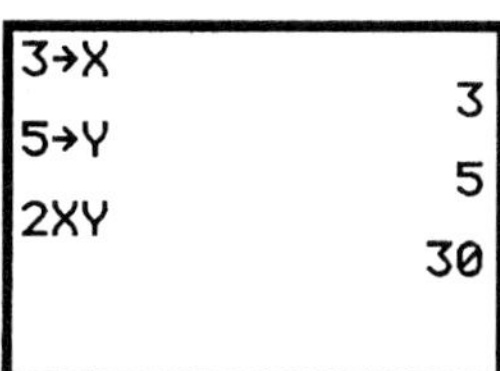

Figure 0.22

There are two ways to access the x variable since it is usually the variable of choice in algebra. Press $\boxed{\text{ALPHA}}$ $\boxed{\text{X}}$ or use the handy $\boxed{\text{X T} \Theta}$ key (see Fig. 0.22).

0.14 Subtraction and "Negative of"

In algebra the minus sign is used two different ways:

1. as the operation sign between two numbers to mean "subtract", as in 5 - 3 or,
2. in front of a number to mean "the opposite of or negative of", as in -7.

The calculator has two different keys for minus. Press 5 $\boxed{-}$ 3 $\boxed{\text{ENTER}}$ for subtraction.

For ‾7 find the negative key $\boxed{(-)}$ located to the left of ENTER . Press $\boxed{(-)}$ 7 $\boxed{\text{ENTER}}$ (see Fig. 0.23).

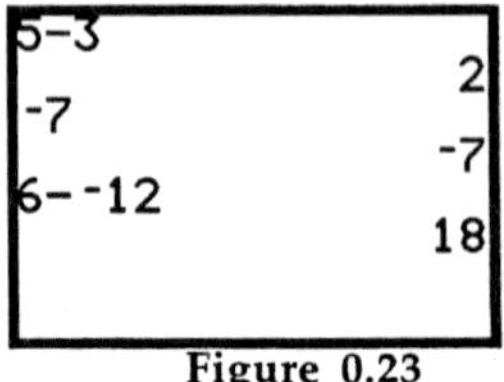

Figure 0.23

Note: The negative sign is actually a little bit shorter and slightly raised compared to the subtraction symbol.

Example 9
Type the following problems:
1. 6 - ‾ 12
2. - 3 x - 9
3. (-5)² ⟶ $(-5)^2$
4. - 5² ⟶ $- 5^2$

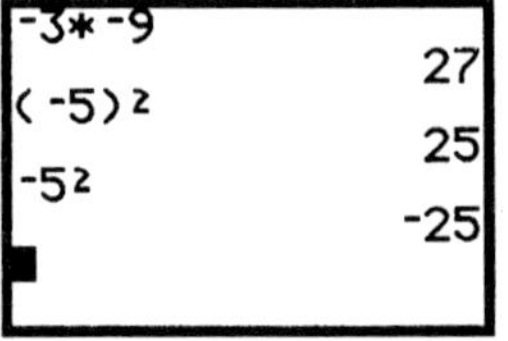

Figure 0.24

(see Figs. 0.23 and 0.24).
Note that the values for example 9 **parts 3 and 4** above are different (see Fig. 0.24.) Order of operations in **part 4** says: "Square five then take its opposite."

Note: To square a negative number you must put it in parentheses.

Trouble Shooting: The most common calculator error is using the subtraction symbol instead of the negative symbol (see Fig. 0. 25 and 0.26).

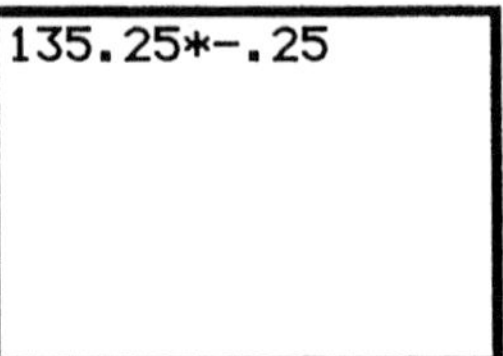

Figure 0.25

0.15 The Error Message

Using the subtraction sign incorrectly produces an error message. When you type the expression as in Fig. 0.25 and press $\boxed{\text{ENTER}}$ **ERR:SYNTAX** appears(see Fig. 0.26).

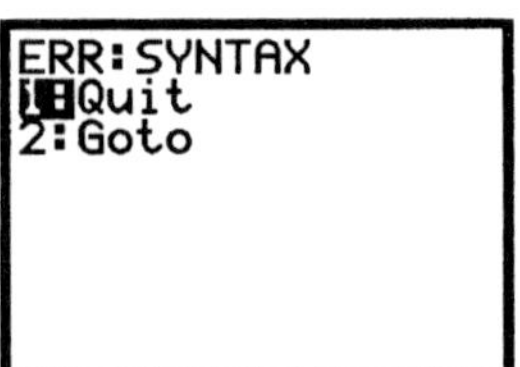

Figure 0.26

Choose [2:Goto] to position the cursor to the place where the error occurred. Choose [1:Quit] to begin a new line on the Home Screen (see Fig. 0.26).

Chapter 1
Visualizing Data Using Statistics Menus

1.1   Using Lists for Data Entry

To enter data into the calculator you use the statistics menu. You can store data into lists labeled L1 through L6.

> **Note:** The TI-83 can store data to additional lists by giving them a name. See the owners manual for more detail.

Press ⎯STAT⎯ . To clear any data stored in List 1, and List 2, select [4:ClrList] then press ⎯2nd⎯ ⎯L1⎯ ⎯,⎯ ⎯2nd⎯ ⎯L2⎯ ⎯ENTER⎯ (see Figs. 1.1 and 1.2).
Now you are ready for data entry.

Press ⎯STAT⎯ ; select [1:Edit] (see Fig. 1.3).

Example 1
Ten students were surveyed. Make a histogram for the number of hours the students worked last week :

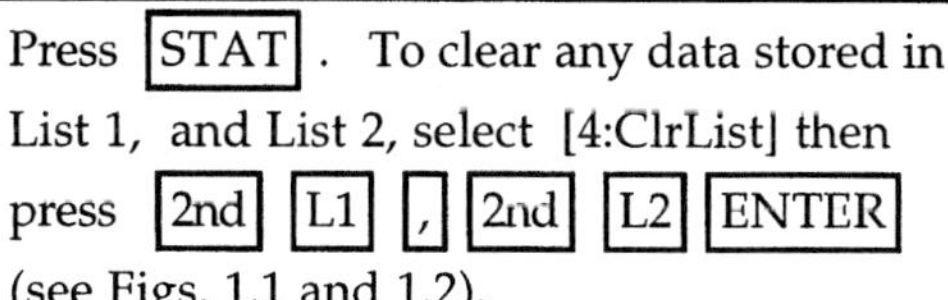

hours worked
0
0
4
10
10
16
16
18
20
20

Table 1. 1

Enter the above data into List 1 (L1).
In column L1 type ⎯0⎯ ⎯ENTER⎯ ⎯0⎯ ⎯ENTER⎯ ⎯4⎯ ⎯ENTER⎯ ... etc. Always press ⎯ENTER⎯ or ⎯∇⎯ after each data entry (see Fig. 1.4).

> **Trouble Shooting:** Placing the cursor on the label L1 will display the list using braces { }. If you press ⎯CLEAR⎯ the list will be deleted. This is a quick way to clear. When the down arrow is pressed you can begin entering data into the list.

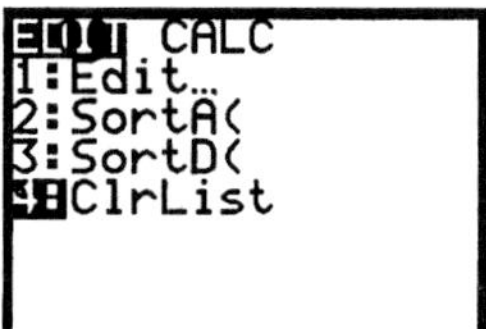

Figure 1. 1

Figure 1. 2

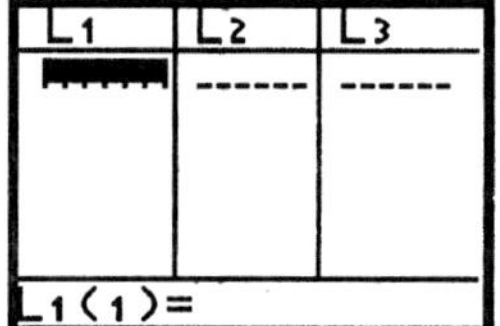

Figure 1. 3

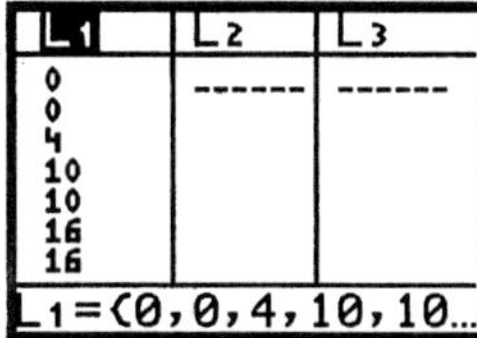

Figure 1. 4

1. 2   Histogram Setup

A histogram is a graph that helps you visualize data. To graph the data in Table 1.1, you must first set up the statistical plot.

Clear all old plots. Press 2nd STAT PLOT select [4:PlotsOff] ENTER . Press 2nd STAT PLOT ; select [1:Plot1] as shown in Figure 1.5. To set up the histogram correctly see Figure 1.6 or 1.7 or follow the instructions below:

1.   Select **ON**. Press ENTER .

2.   Select the graph **Type:** ∇ ▷ ...to the **histogram icon**. Press ENTER .

3.   Select the **Xlist: L1** . Press ENTER (for your horizontal axis).

Note: For the TI-83, press 2nd L1 .

Select the frequency, **Freq**. In Example 1 each item has a frequency of one so select 1

Press ENTER . Your plot should be set up as in Figure 1.6 or 1.7.

Figure 1. 5

Figure 1. 6 TI-82 Plot1 menu.

Figure 1. 7 TI-83 Plot1 menu,

1.3   Selecting the Correct Window for the Histogram

Note: Before you plot a STAT PLOT graph ,

1.   Clear **Y =** or turn off all graphs. To turn off graphs, place the cursor on the = sign then press ENTER .

2.   Turn OFF all plots except the one you want to see.

ERR: INVALID DIM means your lists are not the same size, you have selected a list with no data in it or you have a plot turned on that you did not want and it has different size lists.

1.3.1   Manual Window Sizing

To see your histogram you must set the graphing calculator window to the correct size. Manually set your window size based on the data in L1. Since the L1 data from Example 1 starts at 0 and ends at 20, your x **minimum** and x **maximum** values must be at least: 0 through 20. X maximum should be slightly larger than the highest data value. Y maximum should be slightly larger than the highest frequency of any data entry. Setting the y boundary values at -3 to 5 allows you to see the x-axis as well as the largest frequency value.

Press $\boxed{\text{WINDOW}}$ $\boxed{\nabla}$ and enter the window settings as in Figure 1.8.

Press $\boxed{\text{GRAPH}}$ then $\boxed{\text{TRACE}}$. Use right arrow $\boxed{\triangleright}$ until your screen looks like Figure 1.9.

When you TRACE on a histogram, the cursor moves to the top center of each column. The P1 in the upper right corner indicates you are tracing Plot 1. The **min** = 16, **max** = 1 7 indicates x is on the interval 16 to 17 or, in this case 16 hours. The **n** value (frequency) is 2. So there are two people who worked 16 hours last week.

1.4 Working with More than One List

Example 2
Graph a histogram with two data lists.

List1 represents the sum of the numbers on a pair of dice. List2 represents the possible ways of getting each number by tossing a pair of dice (the frequency). Since we already have data in L1 , enter the data from Table 1.2 below into List 2 (L2) and List 3 (L3).

Clear any data stored in List 2, and List 3. Press $\boxed{\text{STAT}}$; select [4:ClrList]. Press $\boxed{\text{2nd}}$ $\boxed{\text{L2}}$ $\boxed{,}$ $\boxed{\text{2nd}}$ $\boxed{\text{L3}}$ $\boxed{\text{ENTER}}$ (see Figs. 1.10 and 1.11).

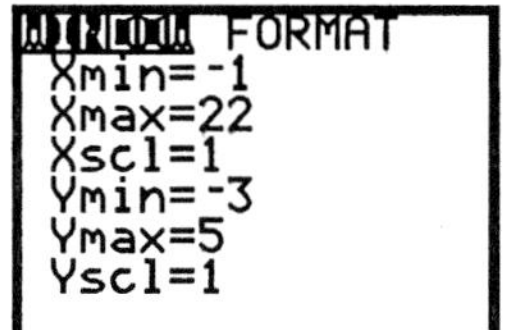

Figure 1. 8

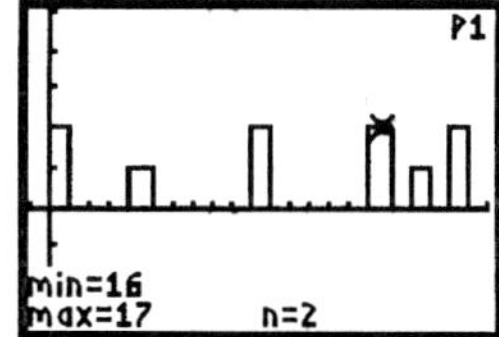

Figure 1. 9

Figure 1. 10

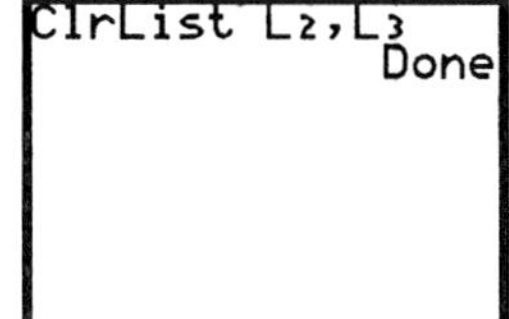

Figure 1. 11

L2	L3
2	1
3	2
4	3
5	4
6	5
7	6
8	5
9	4
10	3
11	2
12	1

Table 1. 2

<u>**Trouble Shooting**</u>: When using lists as frequency, the data must be whole numbers. Decimal values for frequency will produce the error message **ERR:STAT** on the TI-82.

1.4.1   Entering the Data

Press $\boxed{\text{STAT}}$; select [1:Edit] (see Fig. 1.12).
Type in the L2 data from Table 2. Press 2
$\boxed{\text{ENTER}}$ 3 $\boxed{\text{ENTER}}$ 4 $\boxed{\text{ENTER}}$ ... etc.;
always press $\boxed{\text{ENTER}}$ or $\boxed{\text{V}}$ after each data
entry (see Fig. 1.13).

Press $\boxed{\triangleright}$ to get to column L3. Type in all
the L3 data from Table 1.2.

Press $\boxed{\text{ENTER}}$ or $\boxed{\text{V}}$ after each number (see
Fig. 1.13).

1.4.2   Turn Off Old Plots

Press $\boxed{\text{2nd}}$ $\boxed{\text{STAT PLOT}}$; select [4: PlotsOff]
then $\boxed{\text{ENTER}}$ (see Figs. 1.14 and 1.15).

1.4.3   Setting up Plot2

Press $\boxed{\text{2nd}}$ $\boxed{\text{STAT PLOT}}$; select [2: Plot2].

1. Select **ON** press $\boxed{\text{ENTER}}$.

2. Select the graph Type: **histogram icon** .
   Press $\boxed{\text{ENTER}}$.

3. Select the Xlist **L2** (for your horizontal
   axis). Press $\boxed{\text{ENTER}}$.

4. Select the frequency **L3**. Press $\boxed{\text{ENTER}}$
   In Example 2 each item has a frequency
   located in L3.

Set up the plot as in Figure 1.16 or 1.17.

Figure 1. 12

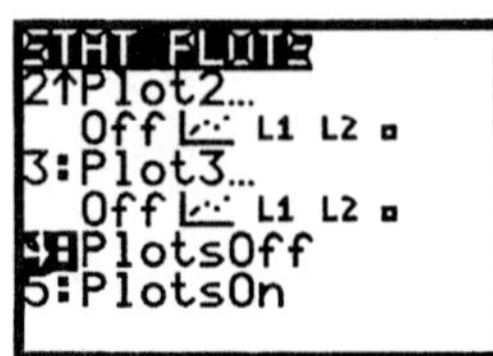

Figure 1. 13

Figure 1. 14

Figure 1. 15

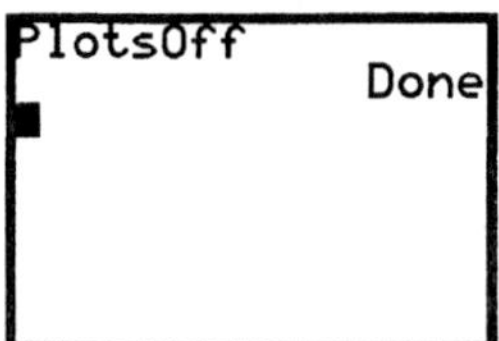

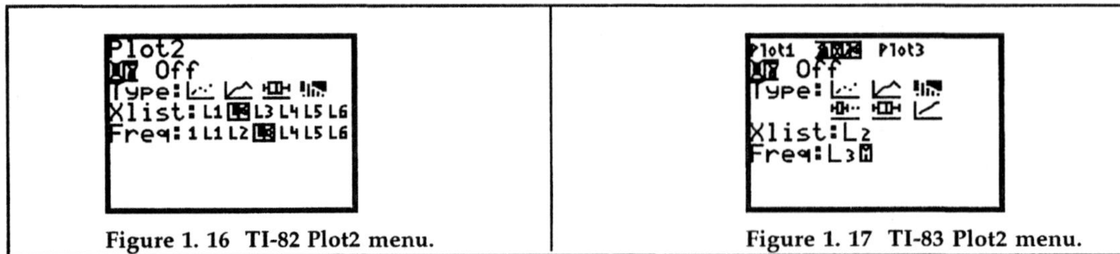

Figure 1. 16 TI-82 Plot2 menu.   Figure 1. 17 TI-83 Plot2 menu.

1.4.4   Size the Window

Size the window to accomodate the data in
Table 1.2 . Press $\boxed{\text{WINDOW}}$ $\boxed{\text{V}}$; then enter
the window settings as in Figure 1.18.

Press $\boxed{\text{GRAPH}}$.

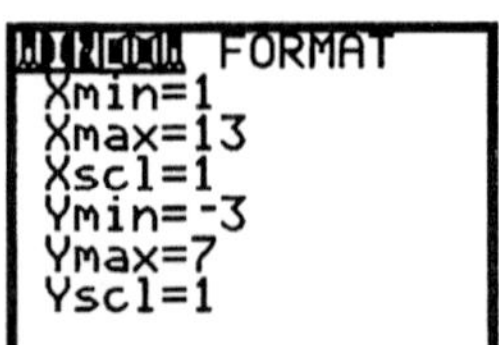

Figure 1. 18

Note: **Automatic Window Sizing.** You can have the calculator automatically set your window by pressing ZOOM ,then select [9:ZoomStat] (see Figure 1.19). You lose control of the max and min values, however, and you may still want to adjust the window.

Figure 1. 19

1.4.5　Tracing on the Histogram

Press GRAPH then TRACE and use your right arrow ▷ until your screen looks like Figure 1.20 or 1.21 . When you TRACE on a histogram, the cursor moves to the top center of each interval column. Your histogram represents the possible ways of getting each outcome by tossing a pair of dice. The P2 tells that you are tracing on Plot2. The min = 6, max = 7 indicates that you are tracing on the interval 6 to 7 or, in this case, 6. The **n** value (frequency) is 5. This means there are five possible ways to get the number six when a pair of dice is thrown.

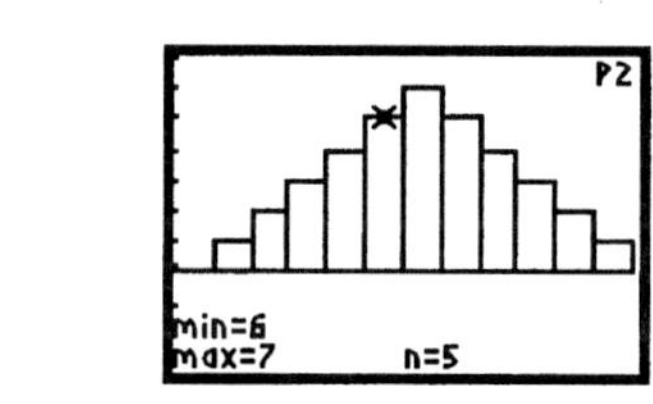

Figure 1. 20 TI-82 Plot2.

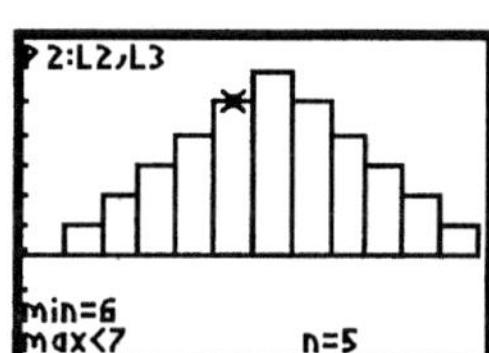

Figure 1. 21 TI-83 Plot2.

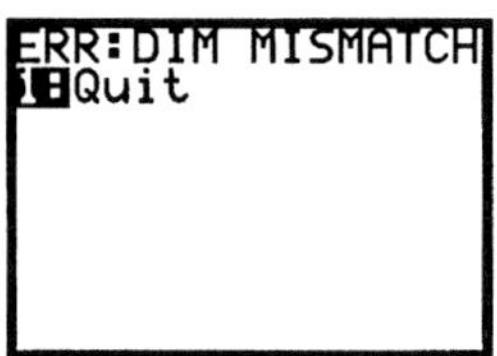

Trouble Shooting: **Dimension Mismatch.** The error message shown in Figure 1.22 is telling you that your lists are not the same size. Either one list is longer than the other or you have set up your plot with the wrong lists. To graph two lists they must have the same number of elements.

Figure 1. 22

1.5　Changing the Interval Width.

Change your WINDOW to the settings as in Figure 1.23. When you change the **Xscl**, you are setting the width size of the interval. In this case the histogram will be 2 units wide, beginning at **Xmin**.

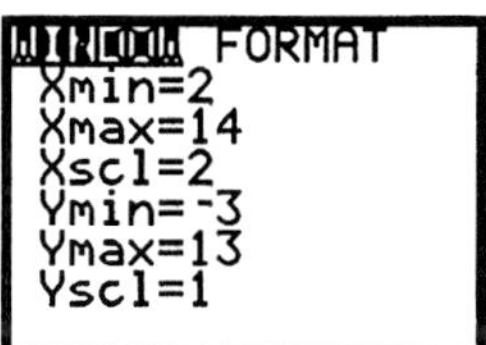

Figure 1. 23

Press GRAPH . Figure 1.24 shows the new histogram with interval width of 2 units. You would interpret the TRACE point to mean there are 11 ways to get a 6 or 7 on the roll of two dice.

Change your WINDOW so that **Xscl** = 3 and **Ymax** = 16.

Figure 1.25 shows the histogram with an interval width of 3 units. Press TRACE ▷ The second interval has n = 15. This means there are 15 ways to get a 5,6, or 7 on a roll of two dice.

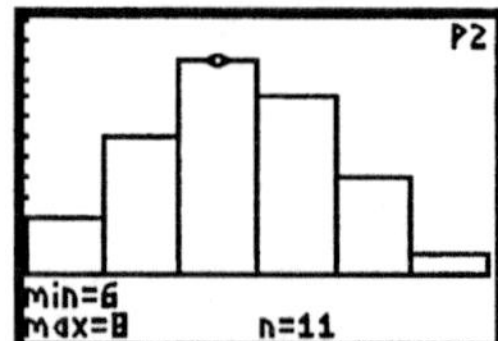

Figure 1. 24

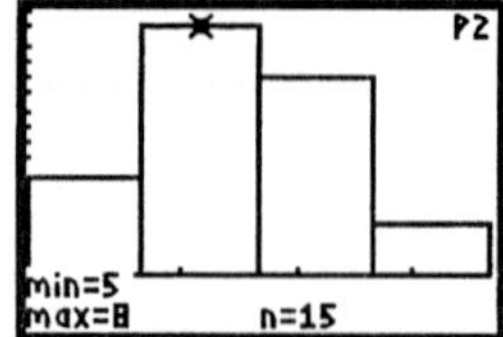

Figure 1. 25

1.6   One Variable Statistics
Mean and Median

Example 3
Below are SAT scores of 10 randomly selected students.

SAT	600	640	430	500	510	530	550	370	500	530

Table 1. 3

Enter this data using the STAT menu. First **CLEAR** all values from L1, L2, and L3 using the techniques learned in Section 1.1 and 1.4 above (see Fig. 1. 26).

Enter the SAT scores in L1. Refer to Section 1.1 if you need help. Press ENTER after each number (see Fig. 1.27).

1.6.1   One Variable Statistics.
Let us look at the *one-variable* statistics performed on List1 (L1).

Press: STAT ▷ to < CALC>. Select [1:1-Var Stats] (see Fig. 1.28).

Then press 2nd L1 ENTER (see Fig. 1.29).

Note: To perform *one-variable* statistics on L2 or any other list, select [1:1-Var Stats] followed by the list number or name.

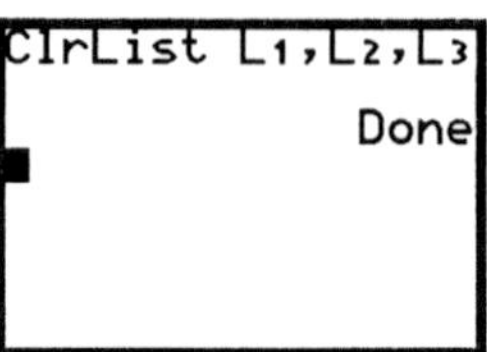

Figure 1. 26

Figure 1. 27

Figure 1. 28

We see the command line in Figure 1.29; press $\boxed{\text{ENTER}}$ to see the L1 *one-variable* statistics information in Figure 1.30.

1.6.2   The Mean.
In Figure 1.30 the statistics we are most interested in are:
1. $\bar{x}$, the **mean** (average) represented by $\bar{x} = 516$ and
2.   **n** , the number of elements in the list , **n = 10**,

This means that 516 was the average (mean) SAT score for the ten students.

1.6.3   The Median
Press $\boxed{\nabla}$ to see the information in Figure 1.31. We are interested in **Med**, the **median** score. The median, represented by **Med = 520**, is the middle score when the data is in rank order. This means that of the ten SAT scores, five will be above 520 and five SAT scores will be below 520.

1.6.4   Sorting a List.
Sort L1 from low to high. Press $\boxed{\text{STAT}}$, select [2:SortA(] (see Fig. 1.32). Type the list you want to sort. Press $\boxed{\text{2nd}}$ $\boxed{\text{L1}}$ $\boxed{)}$

$\boxed{\text{ENTER}}$ (see Fig. 1.33). Figure 1. 34 shows that we have sorted L1 .

> **Trouble Shooting:** If you have two lists L1 and L2 that you want to sort as an ordered pair, type **SortA(L1,L2)** $\boxed{\text{ENTER}}$. L1 will be sorted in ascending order with L2 as its paired list.

1.6.5   Looking for the Median.
Press $\boxed{\text{STAT}}$; select [1:Edit].

Look at your list now (see Fig. 134). The **median** score is the middle score on the list, however, since there are 10 elements (an even number) the median is the score in between the fifth and sixth score, or
$$\frac{510 + 530}{2} = 520$$

> **Note**: You may think of the median of an even number of elements as the average of the middle two scores.

Figure 1. 29

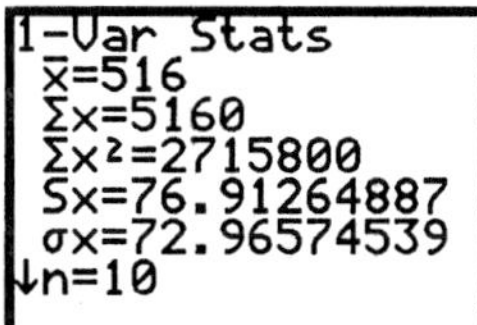

Figure 1. 30

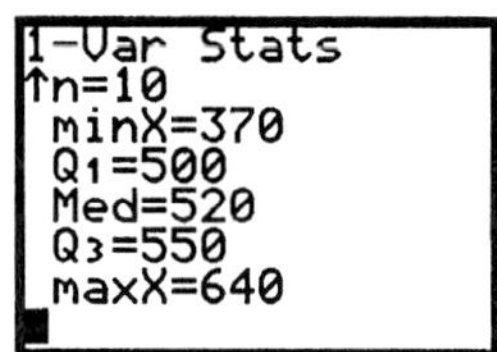

Figure 1. 31

Figure 1. 32

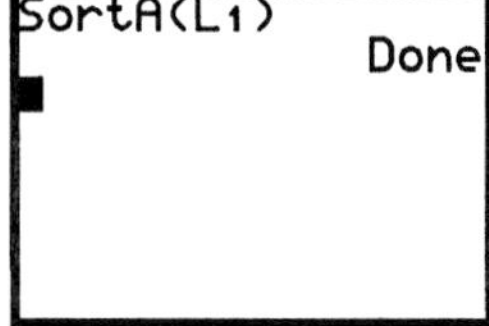

Figure 1. 33

L1	L2	L3
370	------	------
430		
500		
500		
510		
530		
530		

L1(5)=510

Figure 1. 34

1.7 Visualizing the Median; The Box and Whiskers Plot

Do a box and whisker plot. Select a plot. Turn all plots OFF (see Section 1.4.2).

Press [2nd] [STATPLOT] ; select [1:Plot1].

1. Select: **ON** [ENTER] [∇] .

2. Select graph type: [▷] to the **Boxplot icon** [ENTER] (see Fig. 1.35).

Note: TI-83 icons are slightly different.

3. Select Xlist: **L1** [ENTER] .

4. Select Frequency: Choose **1** as the frequency (see Fig. 1. 35).

To graph the Boxplot: press [ZOOM] ; select [9:ZoomStat] (see Fig. 1.36).
The Boxplot shows one-variable statistics.

Press [TRACE] . The middle of the box is the median (**Med**) (see Fig. 1.37). The whiskers on the plot extend from the minimum list value on the left to the first quartile (Q_1) and from the third quartile (Q_3) to the maximum list value. Use [◁] and [▷] arrows to view these scores (see Fig. 1.31 for the Q_1, Med, and Q_3 numerical display).

Note: *Quartiles* are 1/4 or 25% of the list when put in rank order.

1.7.1 Show a Histogram of L1.
Make a histogram. Select the plot.

Press [2nd] [STATPLOT] ; select [1:Plot 1].

1. Select: **ON** [ENTER] .

2. Select: (**histogram icon**) [ENTER] .

3. Xlist: **L1** [ENTER] .

4. Freq: Select **1** (see Fig. 1.38).

Press [WINDOW] [∇] . Select an appropriate range and interval scale based on the data in L1 (see Fig. 1.39).

Press [GRAPH] . Press [TRACE] to view the frequency, **n**, of each SAT score (see Fig. 1. 40).

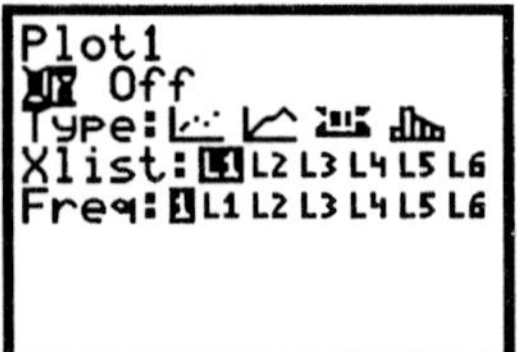

Figure 1. 35

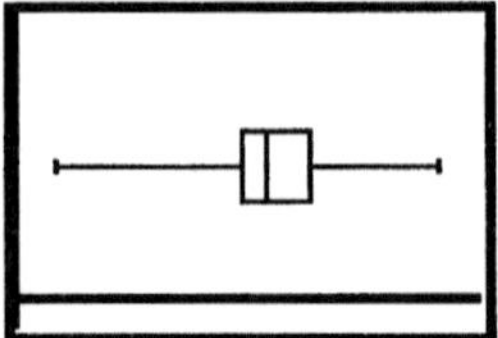

Figure 1. 36

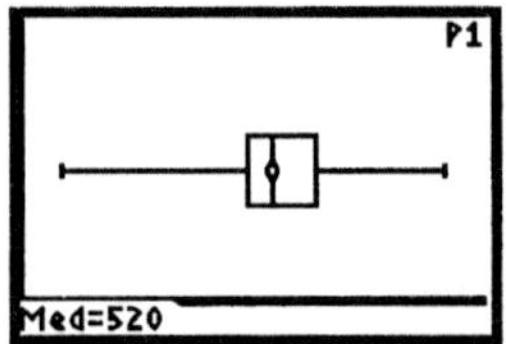

Figure 1. 37

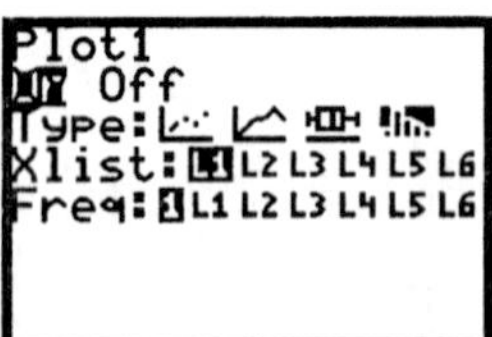

Figure 1. 38

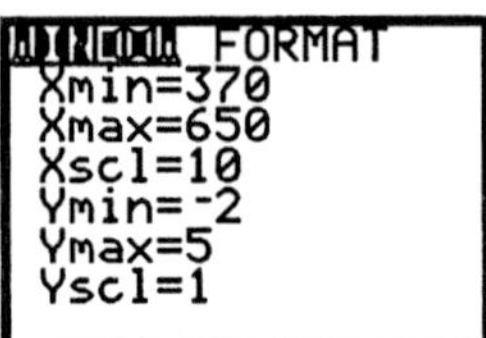

Figure 1. 39

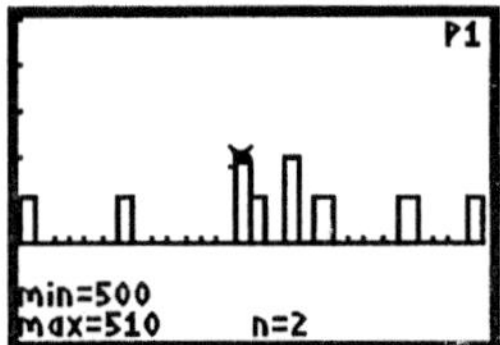

Figure 1. 40

1.8   Other List Techniques

Example 4
Find the mean age of students in a mathematics class.

Age Interval	Frequency Count
15 - 19	2
20 - 24	8
25 - 29	4
30 - 34	3
35 - 39	2
40 - 44	1
45 - 49	1
Total	21

Table 1. 4

Table 1.4 gives the age of students. Since the age is given as an interval, <u>calculate the median of each interval</u> so that a single number can be entered into a list (i.e. 17 is the median of 15 - 19). Go to the home screen and type the median data from Table 1.4 in L1. Remember to enclose the list in braces, { } as in Figure 1.41.

Press [2nd] [{] 17, 22, 27, 32, 37, 42, 47 [2nd] [}]

[STO] [2nd] [L1] [ENTER] . Store the frequency

in L2. Press [2nd] [{] 2, 8, 4, 3, 2, 1, 1 [2nd] [}]

[STO] [2nd] [L2] [ENTER] (see Fig. 1.41).

> **Note**: To calculate the mean you need to multiply the age times the frequency, sum all the ages then divide by the total frequency (see Figs. 1.42 tothrough 1.45).

Give this mean calculation a try from the Home Screen.

1. **Multiply L1 x L2; then store to L3.**

[2nd] [L1] [X] [2nd] [L2] [STO] [2nd] [L3]
(see Fig. 1.42).

2. **Sum L3.**

[2nd] [LIST] [▷] to <MATH>; select [5:sum]

[2nd] [L3] [ENTER] (see Figs. 1.43 and 1. 44).

3. **Divide by the Total Number .**

[÷] 21 (see Fig. 1. 45).

1.8.1   Use the Mean Command for Two Lists

[2nd] [LIST] [▷] to <MATH>, select [3:mean],

[2nd] [L1] [,] [2nd] [L2] (see Fig. 1.45).

Either method gives the same answer.

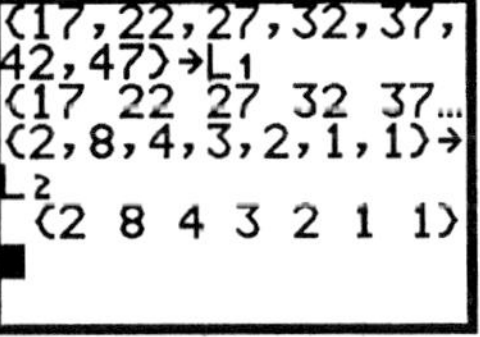

Figure 1. 41

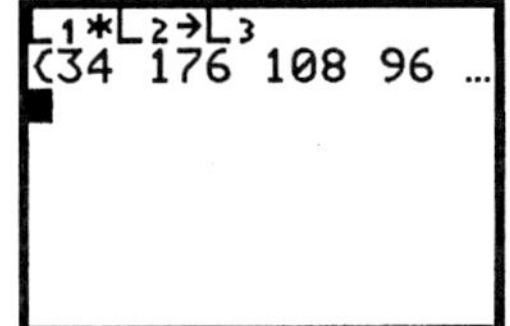

Figure 1. 42

Figure 1. 43

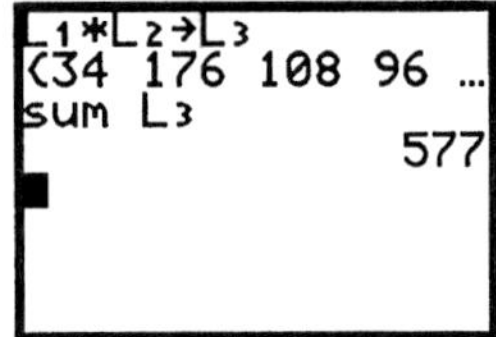

Figure 1. 44

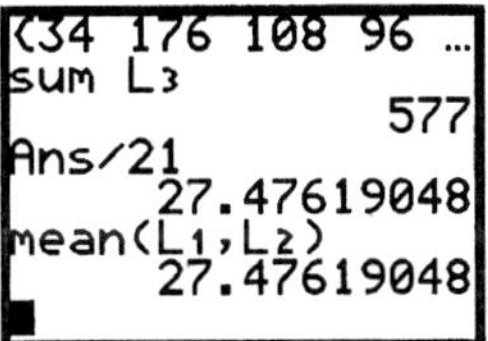

Figure 1. 45

Chapter 2
Scatter Plots and Introduction to Graphing

2.1 Scatter Plots

Relationships between two variables can be visualized by graphing data as a scatter plot. Think of the two list as ordered pairs. An ordered pair (x, y) can represent a point on a graph.

Example 1

In Table 2.1 are the SAT and math placement scores of ten randomly selected freshmen students. Graph the data to see if a relationship exits between the SAT scores and the math placement scores.

SAT	600	640	430	500	510	530	550	370	500	530
Math Place	25	29	14	12	11	8	17	16	9	26

Table 2. 1

> **Note**: Refer back to Chapter 1 Sections 1.1 through 1.4 for information on entering data into lists.

2.1.1 Enter the Data
CLEAR List 1 (L1)and List 2 (L2).
Enter the math placement scores into L1.
Enter the SAT scores into L2 (see Fig. 2.1).

2.1.2 Set Up the Scatter Plot
Press [2nd] [STATPLOT] . Select [1:Plot1].

Set up the Scatter Plot as in Fig. 2. 2 or 2.3.

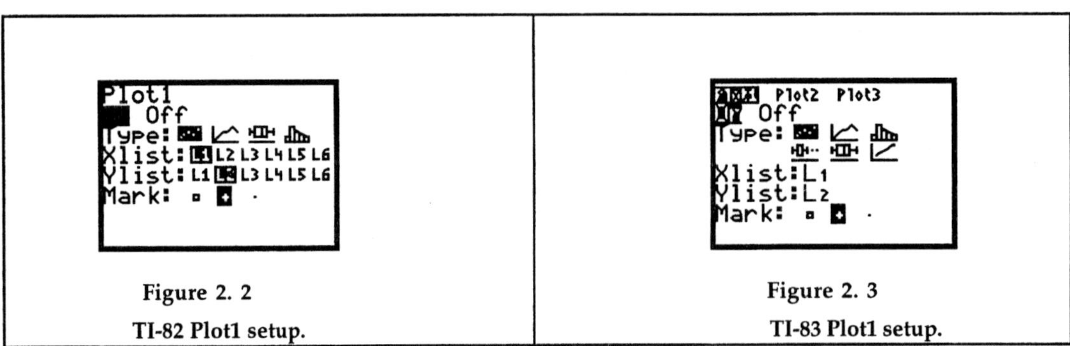

Figure 2. 1

<table>
<tr><td>

Figure 2. 2

TI-82 Plot1 setup.

</td><td>

Figure 2. 3

TI-83 Plot1 setup.

</td></tr>
</table>

2.1.3 Size the Window

Press $\boxed{\text{WINDOW}}$ $\boxed{\nabla}$. Set the appropriate
window based upon the data in L1 and L2,
as in Figure 2.4.
The **X list** contains the math placement
scores and the **Ylist** contains the SAT scores,
so the window is set just beyond the lowest
and highest data values.
X: [Xmin, Xmax] = X: [6, 32] and
Y: [Ymin,Ymax] = Y: [340, 670].

To view the Scatter Plot press $\boxed{\text{GRAPH}}$ (see

Fig. 2.5). There appears to be an increasing
relationship, i.e. as the math placement
score increases the SAT score increases, but
there are a few exceptions (outliners).

2.1.4 Switch the x and y Variables.

What does the plot look like when L2, the
SAT score, is the independent variable?
Set up the plot as in Figure 2.6 or 2.7.

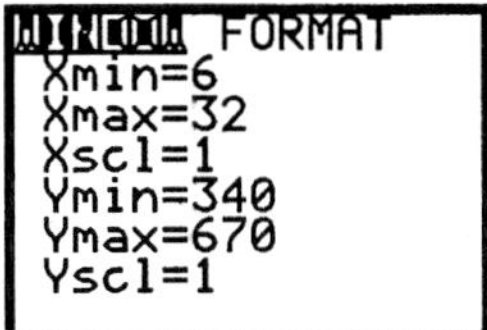

Figure 2. 4

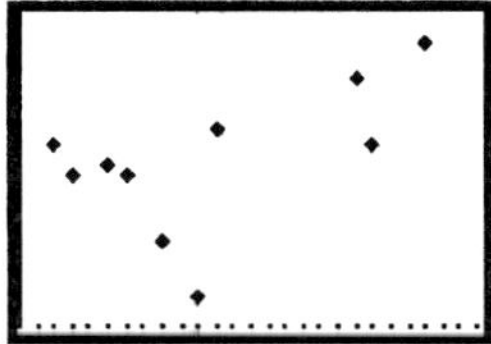

Figure 2. 5

Math placement on the x-axis.

Figure 2. 6 TI-82 Plot1 setup.

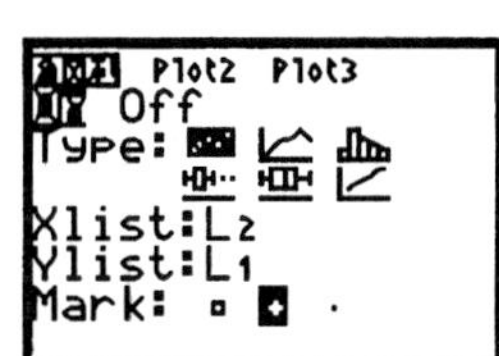

Figure 2. 7 TI-83 Plot1 setup

2.1.5 Sort List 2 and List 1 Together

To sort List2 while keeping the pairing
with List1, Press $\boxed{\text{STAT}}$, select [2:SortA]

$\boxed{\text{2nd}}$ $\boxed{\text{L2}}$ $\boxed{,}$ $\boxed{\text{2nd}}$ $\boxed{\text{L1}}$ $\boxed{)}$ $\boxed{\text{ENTER}}$;
(see Fig. 2.8).

2.1.6 Setting the Window

Look at the sorted list (Fig. 2.9). Press
$\boxed{\text{STAT}}$; select [1:Edit]. Use the sorted List2

to determine the window setting .
Remember we want L2 to be the
independent variable. The window is set
just beyond the lowest and highest data
values. X: [350, 660] and Y: [0, 35].

Press $\boxed{\text{WINDOW}}$ $\boxed{\nabla}$. Set the window as in

Figure 2.10.

Note: There are many correct window sizes.
Choose a window that will contain all of
the points and slightly beyond the points.

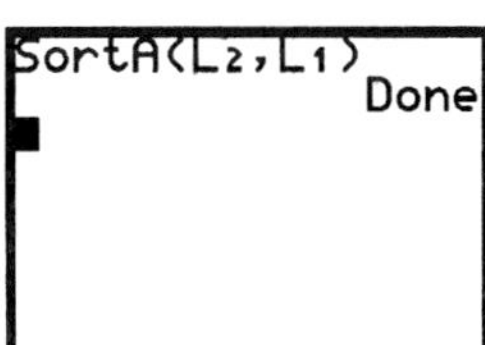

Figure 2. 8

Figure 2. 9

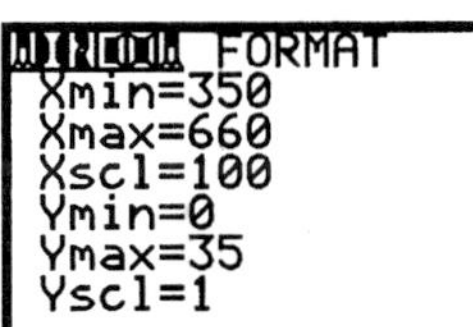

Figure 2. 10

Press GRAPH (see Fig. 2.11). Here the relationship is less clear. Perhaps more data is necessary to determine a trend or relationship. Explore the plot using **TRACE** (see Fig. 2.12).

2.1.7   A Check List for Plotting

To plot statistical data in lists, follow these steps:
1.   **Clear old data in lists.**
2.   **Store the statistical data in one or more lists.**
3.   **Set up the STAT PLOT.**
4.   **Turn Plots ON or OFF as appropriate. (see Fig. 2.13).**
5.   **Clear or deselect [Y=] equations as appropriate (see Fig. 2.14).**
6.   **Define the viewing WINDOW.**
7.   **Explore the plot or graph by pressing TRACE (see Fig. 2.12).**

2.2   Introduction to Graphing Functions

The graphing calculator can be used to graph equations that are functions. The top row of keys, under the viewing screen , contains all the graphing menus.

2.2.1   The Standard WINDOW

Press ZOOM   select [6:Zstandard] (see Fig. 2.15). The graph screen and the $x\,y$ coordinate system appears (see Fig. 2.16).

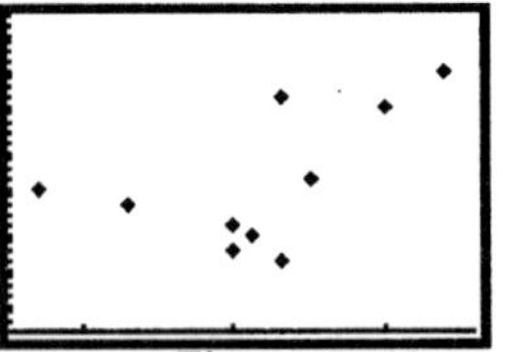

Figure 2. 11
SAT on the *x*-axis

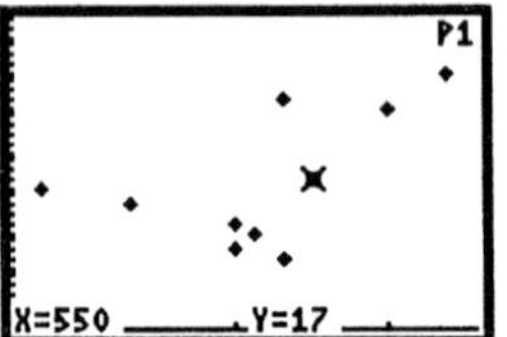

Figure 2. 12

Figure 2. 13

Figure 2. 14

Figure 2. 15

You see only a portion of the real number line on the xy-coordinate plane. The size of the viewing window is determined by the window variables : **Xmin, Xmax, Ymin**, and **Ymax** (see Fig. 2.17).

To see the current *Standard Window* , press WINDOW (see Fig. 2.18).
The distance along the X-axis goes from -10 to 10 or X: [-10, 10]. The distance along the Y-axis goes from -10 to 10 or Y: [-10, 10]. The distance between the tic marks on the *x*-axis is 1 unit (Xscl = 1) and the distance between the tic marks on the *y*-axis is 1 unit (Yscl = 1).

2.2.2   The Free Moving Cursor

Press GRAPH . When you press the arrow keys, $\triangleright$ $\triangleleft$ $\triangle$ $\triangledown$, the cursor can move anywhere on the graphing window. The cursor has changed to a cross and at the bottom of the screen you see the coordinates of the screen position, which change as you jump form pixel to pixel. A pixel is a point of light on the screen (see Fig. 2.19).

2.2.3   The Decimal Window

Notice that on the standard window you get rather "ugly" decimals when you press the arrow keys.

Press ZOOM ; select [4:ZDecimal].

Now press the arrow keys. As you jump from pixel to pixel you increment by the decimal value 0.1 (1/10) (see Fig. 2.20).

Press WINDOW to see the decimal window settings for *x* and *y* (see Fig. 2.21).

Example 2
To change Centigrade temperature to Fahrenheit use the formula F = (9/5)C +32.
Enter this formula into the calculator and graph the function.

Press Y= ; into **Y1** type **(9/5)** *x* **+ 32**.
(see Fig. 2.22 or 2.23).

Trouble Shooting: The independent variable, in this case C, must be entered as X on the calculator. The dependent variable F will become Y. Notice you can graph 10 different functions, Y1 to Y10.

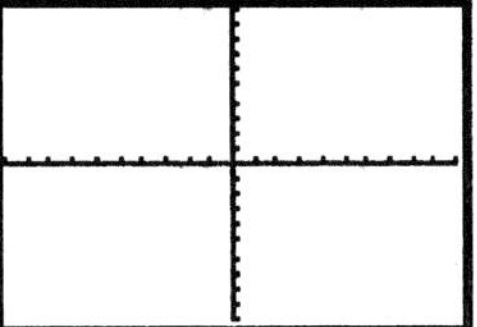

Figure 2. 16

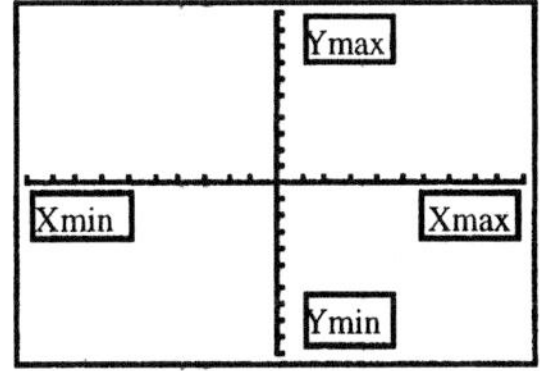

Figure 2. 17
(The words will not appear on your screen)

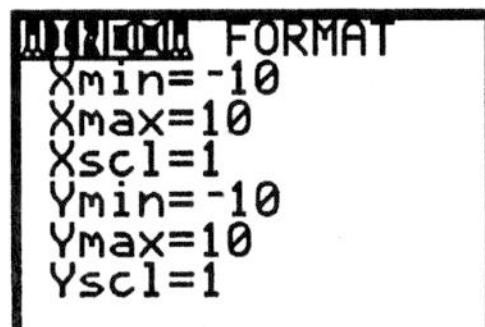

Figure 2. 18

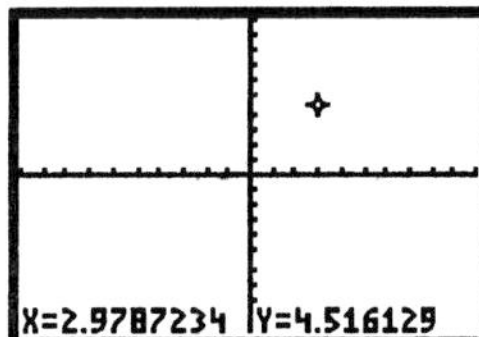

Figure 2. 19

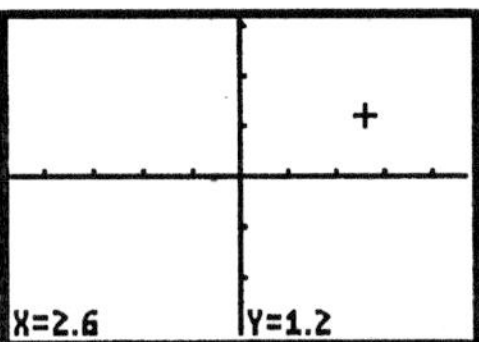

Figure 2. 20

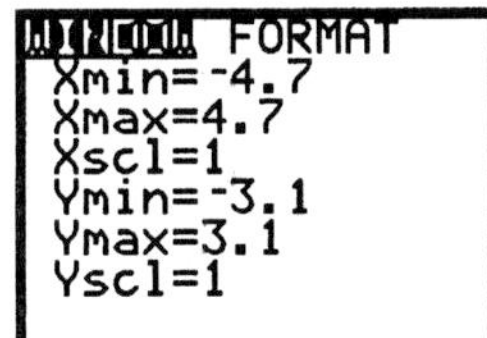

Figure 2. 21

<table>
<tr><td>

```
Y1圖(9/5)X+32
Y2=
Y3=
Y4=
Y5=
Y6=
Y7=
Y8=
```

Figure 2. 22

The TI-82 $\boxed{Y=}$ **Screen**

</td><td>

```
Plot1  Plot2  Plot3
\Y1圖(9/5)X+32
\Y2=
\Y3=
\Y4=
\Y5=
\Y6=
\Y7=
```

Figure 2. 23

The TI-83 $\boxed{Y=}$ **Screen**

</td></tr>
</table>

<u>A Special Note to TI-83 Users.</u>

Plots can be turned **ON** and **OFF** from the $\boxed{Y=}$ screen. To turn a plot ON, use the arrow keys to position the cursor on the desired plot name (Plot1, Plot2, Plot3), then press $\boxed{ENTER}$. To turn the plot OFF press $\boxed{ENTER}$. Darkened plots are ON.

2.2.4   The Integer Window

Press $\boxed{WINDOW}$ $\boxed{\nabla}$, then set the window to the settings as in Figure 2.24.

Press $\boxed{GRAPH}$, then $\boxed{TRACE}$. Right arrow to 8° C to see the equivalent temperature 46.4° F (see Fig. 2.25). Left arrow to -10° C for the equivalent 14° F (see Fig. 2. 26). Notice the pixel jumps are now integers of two units. Play with the arrow keys as you TRACE on the function.

```
WINDOW FORMAT
Xmin=-94
Xmax=94
Xscl=10
Ymin=-62
Ymax=62
Yscl=10
```

Figure 2. 24

Friendly Windows

As you move in an x direction from pixel to pixel there are 94 jumps across the screen. Likewise there are 62 pixel jumps in a y direction. To determine the horizontal jump, Δx, and the vertical jump, Δy, use the following formulas:

$$\Delta x = \frac{x_{max} - x_{min}}{94}$$

$$\Delta y = \frac{y_{max} - y_{min}}{62}$$

Note: A "Friendly Window" is any window whose distance between **Xmax** and **Xmin** is evenly divisible by 94. In the above window $\Delta x = (94 - (-94)) / 94 = 2$, or two units per each x pixel jump.

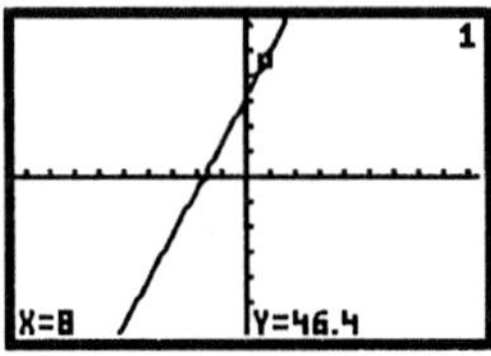

Figure 2. 25

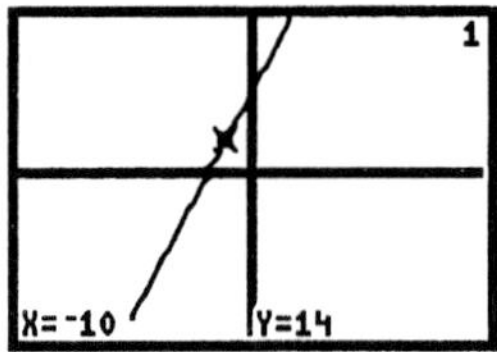

Figure 2. 26

2.2.5 Use Table to Help Find the Appropriate Window.

Finding the appropriate window for a function takes a lot of practice. The following tips can be very helpful:

1. Use algebra to calculate a few easy values of y, such as when $x = 0, 1, -1$ etc.

2. Press $\boxed{\text{TRACE}}$ to discover where a few points lie on the graph.

3. Use $\boxed{\text{2nd}}$ $\boxed{\text{TABLE}}$ to see many values of x and y.

Example 3
Find an appropriate window for
$$f(x) = y = x^2 + 25$$

Press $\boxed{\text{Y=}}$; into **Y1** type: $x^2 + 25$ (see Fig. 2.27). Press $\boxed{\text{ZOOM}}$; select [6:ZStandard]. No graph is seen (see Figs. 2.28 and 2.29).

Algebra tells us that when $x = 0$, $y = 25$. On the standard viewing window the y values are Y: [-10, 10], so $y = 25$ cannot be seen!

Press $\boxed{\text{TRACE}}$; hold down $\boxed{\triangleright}$ to confirm that y is larger than 25 (see Fig. 2.30). Press $\boxed{\text{2nd}}$ $\boxed{\text{TblSet}}$. Begin the table at $x = -5$ and increment by 1 unit (Δtbl = 1). Set up a table of values as in Figure 2.31 . Press $\boxed{\text{2nd}}$ $\boxed{\text{TABLE}}$. Use $\boxed{\text{V}}$ to scroll through some values (see Fig. 2. 32).

An appropriate window based upon the table appears to be X :[-10, 10] and Y: [-10, 50] or better still Y: [-10, 100], with Yscl = 10 (see Fig. 2.33).

Trouble Shooting: To lessen the trial and error process for finding an appropriate viewing window, always use **TRACE** and **TABLE** to find points on your graph. Usually it is the Ymax that needs to be adjusted not Xmax.

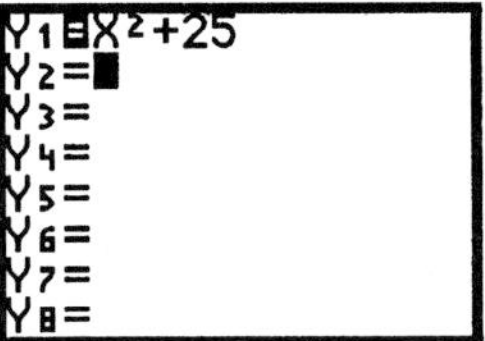

Figure 2. 27

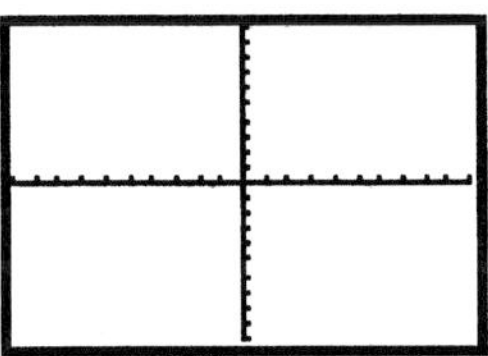

Figure 2. 28

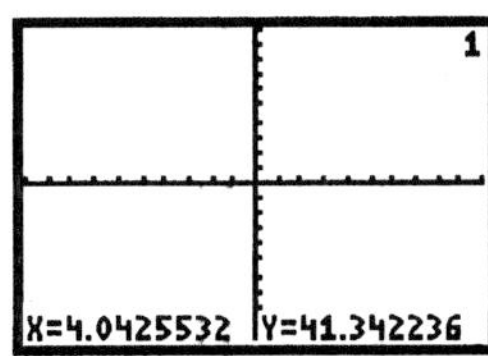

Figure 2. 29

Figure 2. 30
TRACE reveals points on the graph.

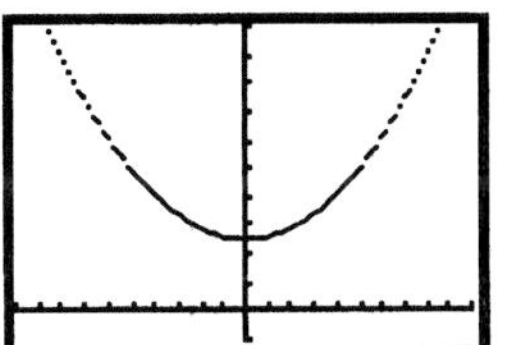

Figure 2. 31

Figure 2. 32

Figure 2. 33
X: [-10, 10] and Y: [-10, 100]

23

2.3 Using Graphs To Determine the Domain of a Function

For most functions the *domain* (possible x values) is all real numbers. However there are functions that are the exception.

Example 4
Use a graph and a table of values to help determine the domain of the function
$$f(x) = y = 1/x.$$

Press $\boxed{Y=}$ $\boxed{CLEAR}$. Into Y1 type $1 / x$

(see Fig. 2.34). Press $\boxed{ZOOM}$; select

[4:ZDecimal] $\boxed{TRACE}$.

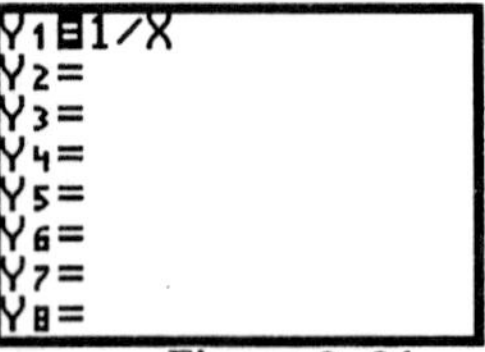
Figure 2. 34

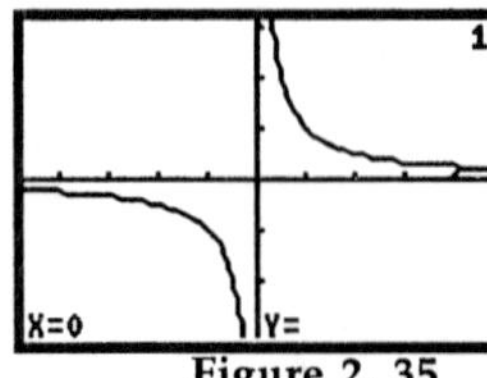
Figure 2. 35
y is undefined for $x = 0$

We see a graph in two pieces. TRACE is telling us that there is no y value associated with $x = 0$ (see Fig. 2. 35).

Set up a table of values. Press $\boxed{2nd}$ $\boxed{TblSet}$. Begin the table at $x = -5$ and increment by 1 unit (Δtbl=1) (see Fig. 2.36). Press $\boxed{2nd}$

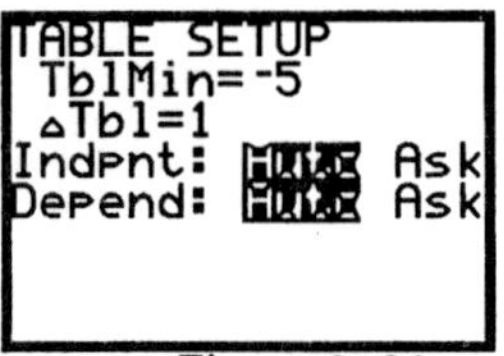
Figure 2. 36

$\boxed{TABLE}$. Notice that for $x = 0$ we get an ERROR message, which means that y is undefined for $x = 0$ (see Fig. 2.37).

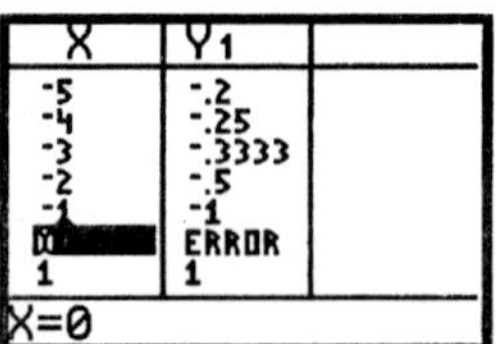
Figure 2. 37

Both the graph and the table confirm that $x = 0$ is not in the domain of x. The domain of $f(x) = 1/x$ is the set of all real numbers x, $x \neq 0$. Using interval notation:
$$(-\infty, 0) \cup (0, +\infty)$$

Example 5
Use a graph and a table of values to help determine the domain of
$$f(x) = y = \sqrt{x}$$

Press $\boxed{Y=}$ $\boxed{CLEAR}$.

Into Y1 type $\boxed{2nd}$ $\boxed{\sqrt{}}$ $\boxed{X,T,\Theta}$.

Press $\boxed{GRAPH}$ $\boxed{TRACE}$.

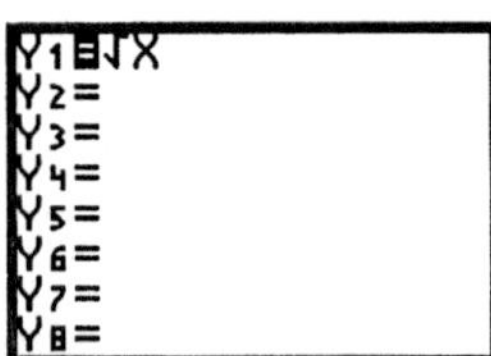
Figure 2. 38

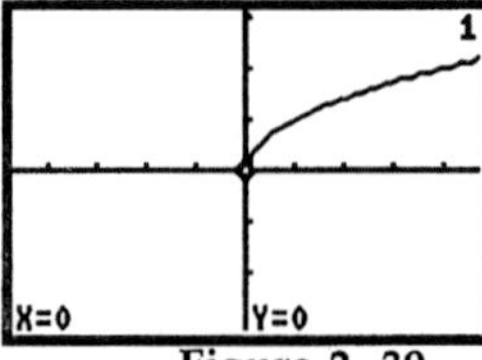
Figure 2. 39

The function $f(x)$ appears to be defined for $x = 0$, but not to the left of zero, $x < 0$ (see Figs. 2.39 and 2.40).

The domain of $f(x) = \sqrt{x}$ is the set of all real numbers $x \geq 0$ or x: $[0, +\infty)$. Confirm with a table of values.

Press $\boxed{\text{2nd}}$ $\boxed{\text{TABLE}}$. Notice that values for x less than 0 give an ERROR message, which means the function is undefined for $x < 0$ (see Fig. 2.41). Use $\boxed{\nabla}$ to see other values of y that are defined by x.

This helps confirms that the domain of the square root function is the set of all x such that $x \geq 0$, i.e., domain $= \{x \mid x \geq 0\}$.

2.3.1 Graphs that are NOT Functions.
Example 6
$$\text{Graph} \quad y^2 = x$$

First solve for y. There are two values for y, $y = +\sqrt{x}$ or $y = -\sqrt{x}$, , therefore y is not a function.   However, a relationship does exist between x and y. We need to trick the calculator into graphing the relationship.

Let **Y1** $= \sqrt{x}$ and **Y2** $= -\sqrt{x}$. (see Fig. 2.42).
Press $\boxed{\text{GRAPH}}$ (see Fig. 2.43).

In addition we can tell y is **not** a function of x because it fails the *vertical line test*.

NOTE: The **Vertical line test** says if a vertical line intersects a graph at more than one point, the graph is not a function.

Press $\boxed{\text{2nd}}$ $\boxed{\text{DRAW}}$; select [4:Vertical] (see Fig. 2.44).

Use the right arrow $\boxed{\triangleright}$ to position the vertical line on $x = 3$.

Since a vertical line intersects the graph in two places (i.e. there are two y values for one x value), the graph is **not** a function (see Fig. 2.45).

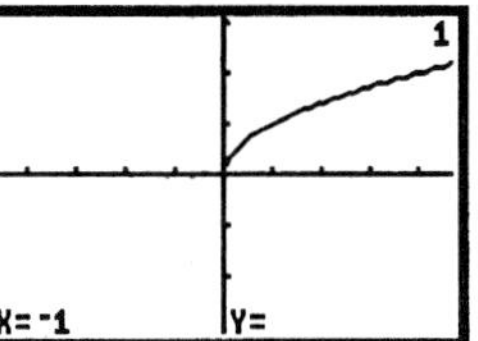

Figure 2. 40

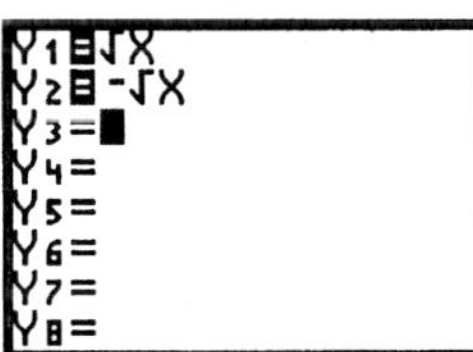

Figure 2. 41

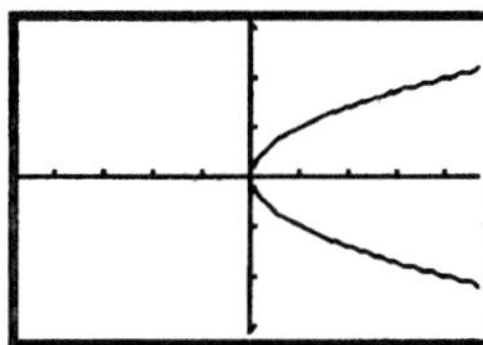

Figure 2. 42

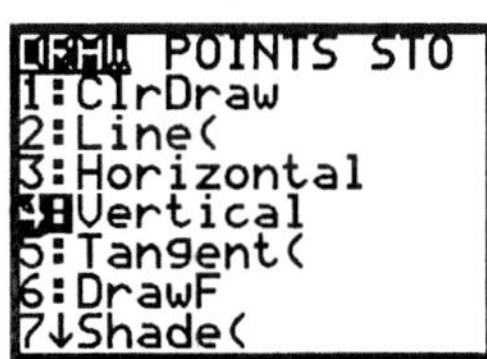

Figure 2. 43

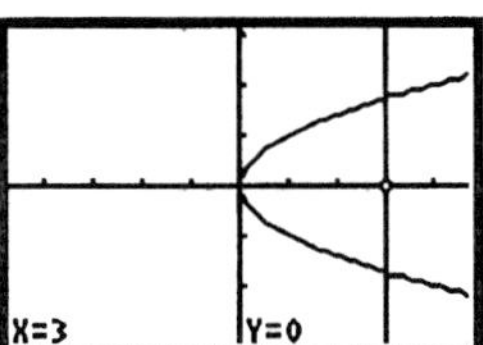

Figure 2. 44

Figure 2. 45

(3, 0) are the coordinates of the x-intercept of the vertical line

2.3.2 Range and Graphs of Other Interesting Functions.

Example 7

1. **Graph the absolute value function**
   $$f(x) = y = |x|$$

Press $\boxed{Y=}$ and $\boxed{CLEAR}$ all values from Y1.

*Type $\boxed{2nd}$ $\boxed{ABS}$ $\boxed{(}$ $\boxed{X,T,\Theta}$ $\boxed{)}$ (see Fig. 2.46)..

> ***TI-83 Note:*** The absolute value command is under $\boxed{MATH}$ <NUM> [1:abs( ]. The left parentheses is provided, you type the right parentheses.

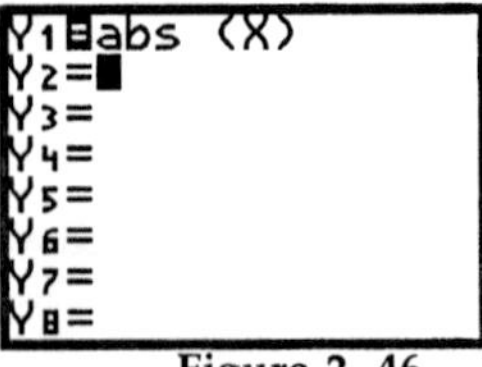

Figure 2. 46

Press $\boxed{ZOOM}$; select [4:ZDecimal].

The absolute value graph has the distinctive "V" shape. The absolute value of a number is a nonnegative value. Therefore, your output values (y values), or the *range* of the function, is $y \geq 0$.

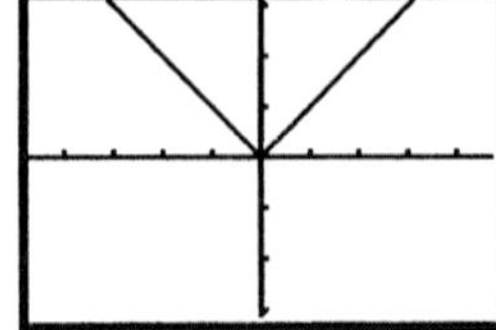

Figure 2. 47

2. **Graph the absolute value function**
   $$g(x) = y = |x + 2| - 1$$

Press $\boxed{Y=}$ into Y2 *type $\boxed{2nd}$ $\boxed{ABS}$ $\boxed{(}$ X + 2 $\boxed{)}$ -1 . (*TI-83 see above box) (see Fig. 2.48).

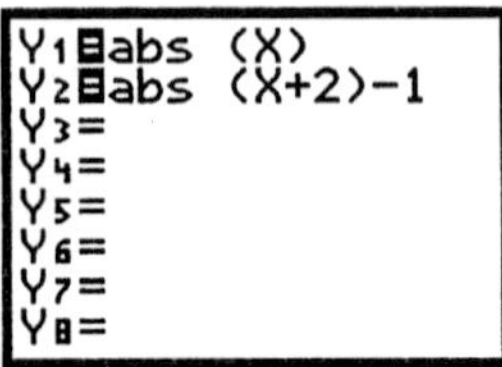

Figure 2. 48

Press $\boxed{GRAPH}$ $\boxed{TRACE}$ $\boxed{\nabla}$. You are on Y2 .
Left arrow to $x = -2$ (see Fig. 2. 49). This reveals the lowest y value on the graph, or $y = -1$. The range for $y = |x + 2| - 1$ is:
$$y \geq -1 \text{ or } \{y \mid y: [-1, +\infty)\}$$

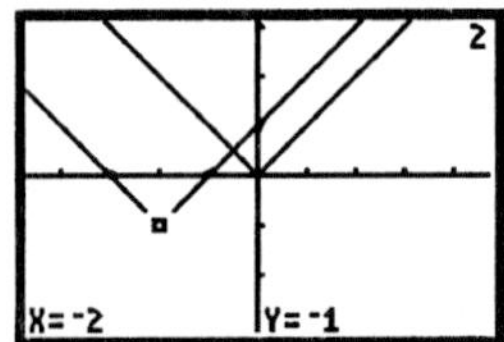

Figure 2. 49

3. **Graph the greatest integer function :**
   $$h(x) = y = [[x]]$$

The **greatest integer function** denoted by $[[x]]$, is defined as the greatest integer less than or equal to x. $[[4.3]] = 4$ and $[[-2.7]] = -3$. This is the interger to the *left* of the number. The graph is often referred to as a step function. To see the graph reset the calculator to **dot mode** (see Fig. 2.50).

Press $\boxed{Y=}$ and $\boxed{CLEAR}$ all expressions.

Figure 2. 50

Press $\boxed{MATH}$ <NUM> select [: int]. Your function should be typed as in Figure 2.51.

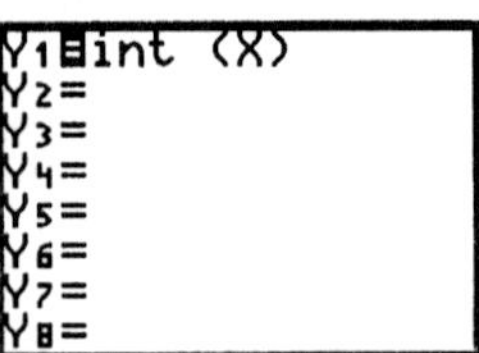

Figure 2. 51

Press $\boxed{GRAPH}$. The distinctive step function appears. The domain for x is any real number, $\boxed{TRACE}$ reveals that the range of y is always an integer (see Fig. 2.52).

> **Note:** Remember to change back to connected MODE.

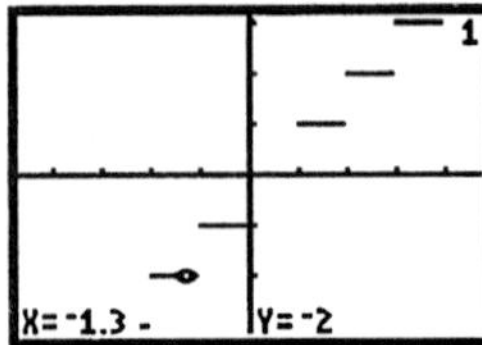

Figure 2. 52

The graph shown in Dot Mode

Chapter 3
Finding Average Rate of Change

3.1 Finding the Average Rate of Change

Example 1

In 1980 the US Federal debt was 909 billion dollars. In 1990 the Federal debt was 3206 billion dollars. Find the average rate of change.

The average rate of change is:

$$\frac{\text{change in debt}}{\text{change in years}} = \frac{3206 - 909}{1990 - 1980} = 229.7$$

or 229.7 billion dollars per year.

This means on average, the US Federal debt increased by \$229.7 billion/yr. from 1980 to 1990 (see Fig. 3.1).

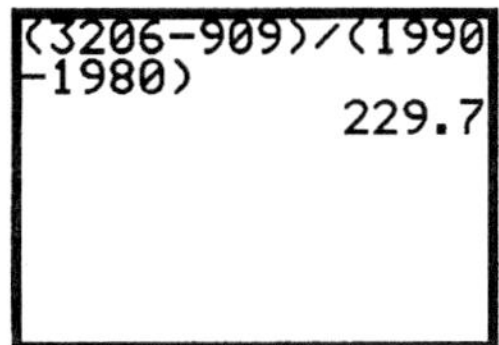

Figure 3. 1

3.1.1 Using Lists to Find the Rate of Change

Enter the data from Example 1 in list 1, L1, and list 2, L2, then calculate the average rate of change.

1. Press $\boxed{\text{STAT}}$; select [4:ClrList];

2. Type: L1, L2, L3 $\boxed{\text{ENTER}}$ (see Fig. 3.2).

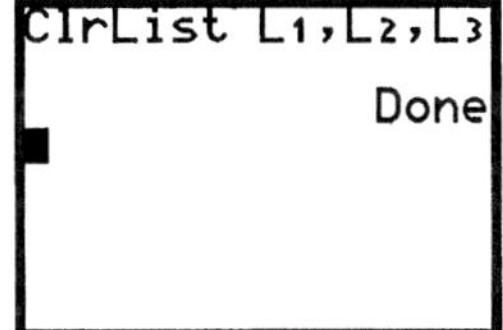

Figure 3. 2

3. Enter the data in L1 and L2. Press $\boxed{\text{STAT}}$; select [1:Edit]. Remember to press $\boxed{\text{ENTER}}$ or $\boxed{\text{V}}$ after every entry (see Fig. 3.3).

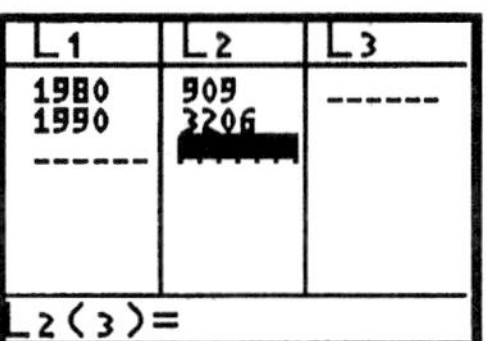

Figure 3. 3

Use the list position to calculate the average rate of change.

Type the following list equation:

$$\frac{(L2(2) - L2(1))}{(L1(2) - L1(1))}$$

Press $\boxed{\text{2nd}}$ $\boxed{\text{QUIT}}$; then type:

($\boxed{\text{2nd}}$ $\boxed{\text{L2}}$ (2) $\boxed{-}$ $\boxed{\text{2nd}}$ $\boxed{\text{L2}}$ (1)) $\boxed{\div}$

($\boxed{\text{2nd}}$ $\boxed{\text{L1}}$ (2) $\boxed{-}$ $\boxed{\text{2nd}}$ $\boxed{\text{L1}}$ (1))

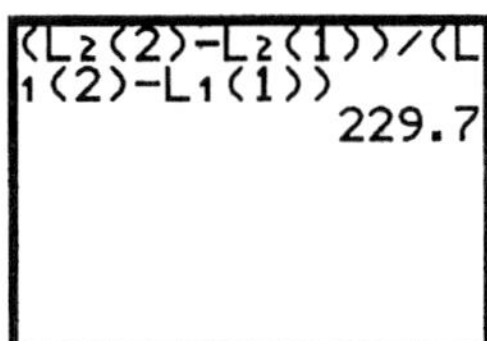

Figure 3. 4
Carefully enter all parentheses.

(see Fig. 3.4). While this may seem tiresome to type, the advantage is that you can change the numbers in the list and then recall the rate of change list equation.

Example 2

In 1985 the federal debt was 1817 billion dollars. Find the average rate of change from 1985 to 1990.

1. Change L1 and L2 to look like Figure 3.5.
2. Recall the rate of change equation on the Home Screen.

Press 2nd QUIT 2nd ENTRY ENTER (see Fig. 3.6).

This means that from 1985 to 1990 the Federal debt increased on average $277.8 billion/yr.

3.1.2 Working With Longer Lists
Example 3

Below is the Federal debt from 1985 to 1990.

Year	Billions of $
1985	1817
1986	2120
1987	2346
1988	2601
1989	2868
1990	3206

Note: As previous noted in section 1.1, use the clear shortcut technique to clear lists:

△ to L1 CLEAR ▽ ▷ △ to L2 CLEAR ▽ .

1. Calculate the average rate of change for each year. Enter the above data into L1 and L2. Press STAT [1:Edit] (see Fig. 3.7).
2. Create the rate of change equation by using the sequence command and store the values to list 3, L3.

The rate of change equation for any year would be :

$$\frac{(L2(N+1) - L2(N))}{(L1(N+1) - L1(N))}$$

Press 2nd QUIT . Type the following:

2nd LIST [1], select [5:seq( ] (see Fig. 3.8).

Type the commands as in Figure 3.9 and store to list 3, L3.

____

[1] For the TI-83 2nd LIST <OPS> select [5:seq( ].

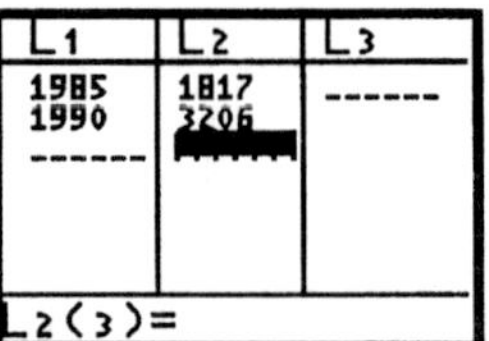

Figure 3. 5

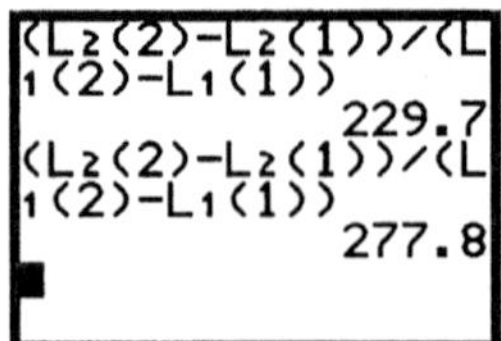

Figure 3. 6

Carefully enter all parentheses.

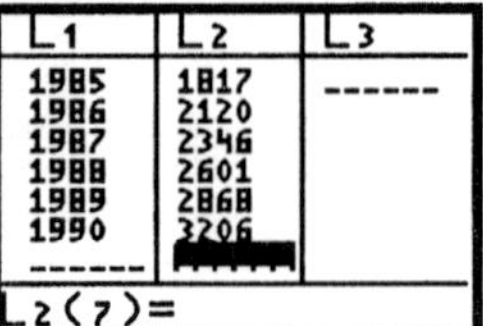

Figure 3. 7

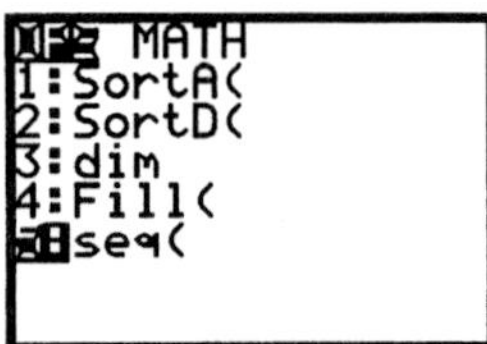

Figure 3. 8

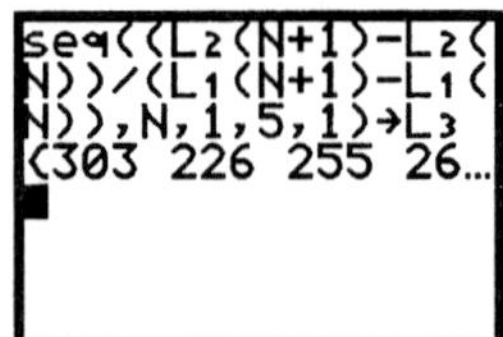

Figure 3. 9

Carefully enter all parentheses

Figure 3.10 shows the rate of change values stored in L3. This is a very handy technique for calculating the rate of change for very long lists.

L1	L2	L3
1985	1817	303
1986	2120	226
1987	2346	255
1988	2601	267
1989	2868	338
1990	3206	------
------	------	

L1(1)=1985

Figure 3. 10

L1	L2	L3
1985	1817	0
1986	2120	303
1987	2346	226
1988	2601	255
1989	2868	267
1990	3206	338
------	------	------

L3(1)=0

Figure 3. 11

Insert zero to make lists the same size.

3.2 Special TI-83 Techniques for Finding the Average Rate of Change of Lists

The TI-83 has a shortcut method for finding the rate of change of a list. Calculate the average rate of change (the difference of elements in L2 divided by the difference of elements in L1), then store it to L3 (see Fig. 3.12).

Press [2nd] [LIST] [▷] to <OPS> , select [7:ΔList( ]. Type as in Figure 3.13.

Press [STAT] [1:Edit] , to see all three lists (see Fig. 314).

See the Trouble Shooting box above Figure 3.11 for graphing the Rate of Change.

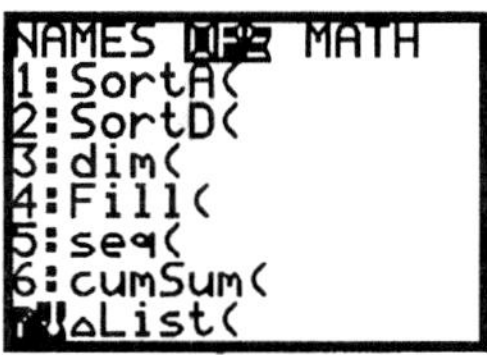

Figure 3. 12 TI-83 Δlist

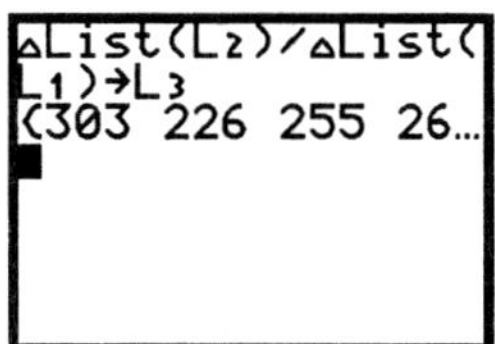

Figure 3. 13 TI-83 Rate of Change

L1	L2	L3
1985	1817	303
1986	2120	226
1987	2346	255
1988	2601	267
1989	2868	338
1990	3206	------
------	------	

L3 ={303,226,255...

Figure 3. 14 TI -83 Lists

Chapter 4
Graphing Linear Functions

4.1   Linear Functions
4.1.1   Definition of a Linear Function
The relationship between the variables x and y is said to be linear if for any set of points (x, y) the rate of change of y with respect to x is constant. The rate of change is called *the slope of the line.* Any collection of points (x, y) that have a linear relationship will satisfy an equation of the form $y = mx + b$ where m is the rate of change of y with respect to x ($m = $ slope) . b is the y - **intercept** (the value of y when $x = 0$). $y = mx + b$ is called the *slope intercept form.*

An example of a linear equation (function) is :
$$y = 5x + 4$$
The slope is: $m = 5$
The y-intercept is: $b = 4$

> **Note**: If needed, refer back to Appendix Section 2.2 to review the introduction to graphing.

Example 1
Plot the following points on an $x\,y$ coordinate plane and determine if they lie on a straight line.
(0, 2), (1, 5), (2, 8), (3, 11), (5, 17), and (7, 23)
Organize the data as in Table 4.1

x	0	1	2	3	5	7
y	2	5	8	11	17	23

Table 4. 1

4.1.2   Plotting Points
You can use the graphing calculator to display the data in list form:
1.   First CLEAR any data in list 1 (L1) and list 2 (L2).

     Press STAT ; select [4:ClrList].

     Press 2nd L1 , 2nd L2 (see Fig. 4.1).

2.   Enter the data points. Press STAT , select [1:Edit] enter the x- coordinate in L1 pressing ENTER after each entry.

     Move the right arrow ▷ to L2 and enter the y-coordinate data (see Fig. 4.2).

3.   Set up the plot. Press 2nd STATPLOT , select [1:Plot 1] as in Figure 4.3 or 4.4.

4.   Clear Y= and turn off all other plots.

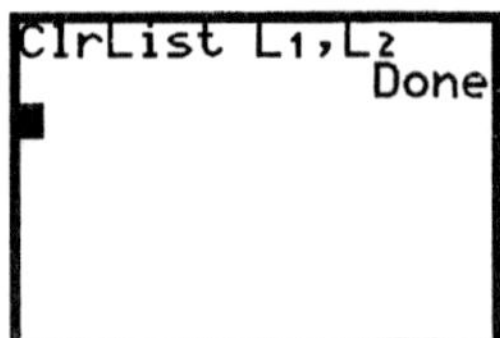

Figure 4. 1

Figure 4. 2

Figure 4. 3   TI-82 Plot 1 screen

Figure 4. 4 TI-82 Plot 1 screen

To see the plot, set the window based upon the highest and the lowest values of x and y . X: [-1,8] and Y: [-2, 25]. Or let the calculator set it:

Press ZOOM ; select [9:ZoomStat] (see Fig. 4.5).

The points appear to lie on a straight line (see Fig. 4.6).

4.1.3   The Algebraic Representation.

You see from the data that when $x = 0$ $y = 2$, so $b = 2$. As x changes one unit, y changes 3 units. The rate of change is 3 units, or $m = 3$. The linear function in slope intercept form $y = mx + b$ is :

$$y = 3x + 2$$

Enter this function into Y1.

Press Y=   CLEAR to clear any old functions.

Press 3 X,T,Θ   + 2 (see Fig. 4.7 or 4.8).

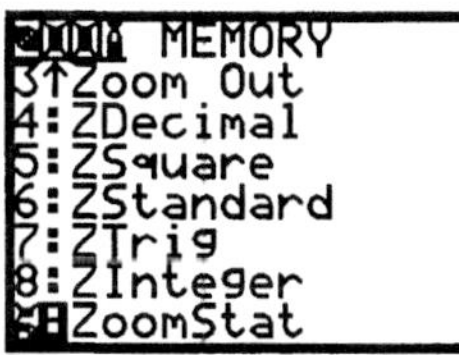

Figure 4. 5

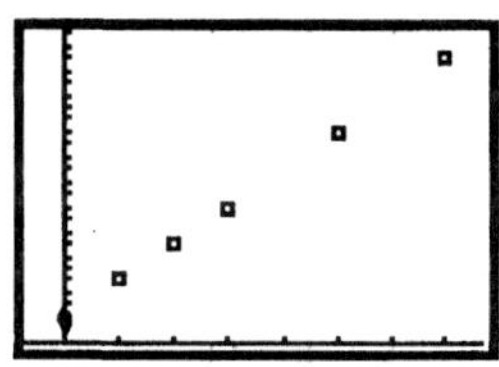

Figure 4. 6

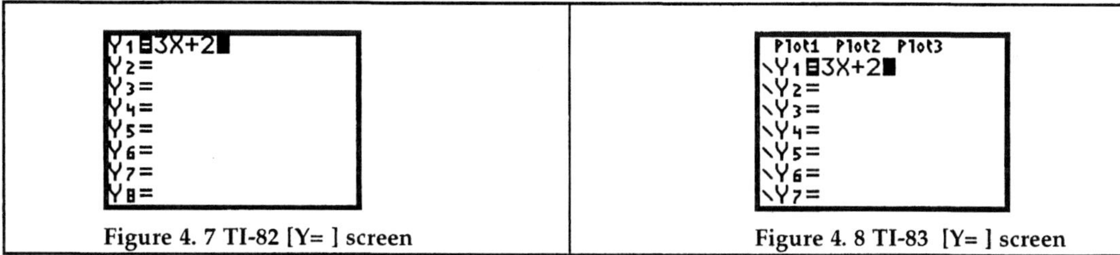

Figure 4. 7 TI-82 [Y=] screen

Figure 4. 8 TI-83 [Y=] screen

Press GRAPH   TRACE (see Fig. 4.9 or 4.10).

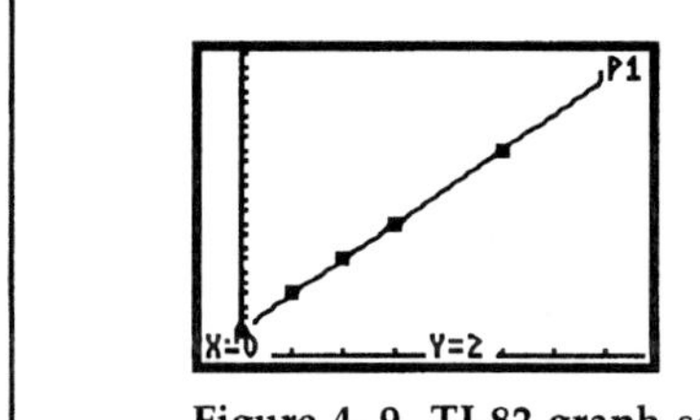

Figure 4. 9 TI-82 graph screen

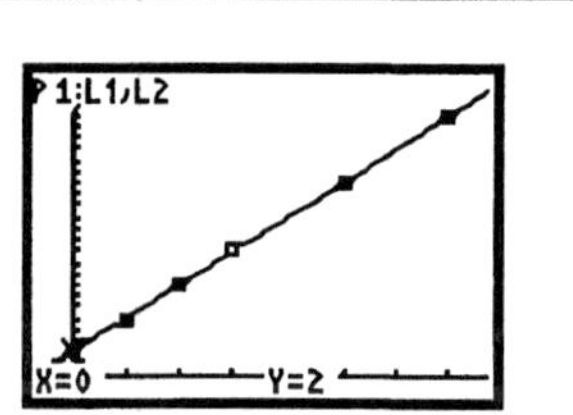

Figure 4. 10 TI-83 graph screen[1]

---

[1] The TI-83 graph screen shows the name of the lists being plotted and the equation. From now on only the TI-82 graph screen will be shown.

Verify that the data points are the same as the points on the graph.

Press $\boxed{2nd}$ $\boxed{TBLSET}$. Set as in Figure 4.11.

Compare the function values of x and y to the data points in Table 4.1.

Press $\boxed{2nd}$ $\boxed{TABLE}$.

Figure 4.12 and Table 4.1 share the same values. The difference is that the algebraic model can be used to find other values of y when x is given. For example use $\boxed{\nabla}$ to find $x = 15$, the corresponding value is $y = 47$ (see Fig. 4.13).
Algebraically;
$$y = 3(15) + 2 = 47$$

Use $\boxed{\Delta}$ to find $x = -10$, the corresponding value is $y = -28$ (see Fig. 4.14).
Algebraically;
$$y = 3(-10) + 2 = -28$$

4.2   Graphing Linear Functions
4.2.1   The ZOOM Menu

The graphing calculator has a default viewing window. This window extends in an x direction from -10 to 10 and in a y direction from -10 to 10.
This is the "standard" window. To set this window quickly;

Press $\boxed{ZOOM}$ $\boxed{6}$ (see Fig. 4.15).

The function and and the plot are displayed.

Press $\boxed{WINDOW}$ to verify the setting (see Fig. 4.16).

Figure 4. 11

Figure 4. 12

Figure 4. 13

Figure 4. 14

Figure 4. 15

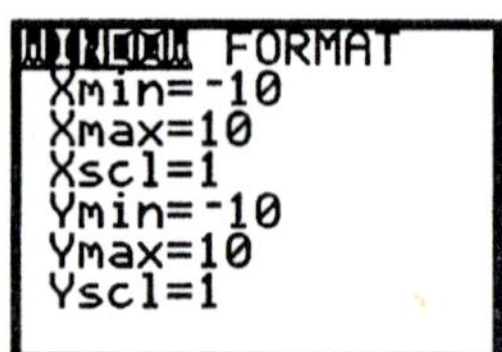

Figure 4. 16

Example 2
Graph the following linear functions in Y1 and Y2:

$$f(x) = x + 5$$
$$g(x) = -2x - 3$$

1. Press $\boxed{Y=}$. $\boxed{CLEAR}$ all the old entries.
   Type the functions (see Fig. 4.17).

2. Press $\boxed{GRAPH}$. Your graphs should be
   the same as Figure 4.18.

Press $\boxed{TRACE}$ then $\boxed{\triangleright}$ $\boxed{\triangleleft}$ to move along the
function. Notice x is being incremented in
rather ugly decimal values. This is NOT a
"friendly" window.

Use $\boxed{\nabla}$ and $\boxed{\Delta}$ to switch between graphs.

Change the window; press $\boxed{ZOOM}$ $\boxed{8}$; the
"integer" window.

> **Trouble Shooting:** You are now given an
> opportunity to reposition your cursor, if you
> should like, before the window changes.
> Press $\boxed{ENTER}$ to activate the change.

If you didn't move your cursor, your graph
will look like Figure 4.20. These are the
graphs of the same functions but the
window is now much larger. Press $\boxed{TRACE}$
and notice that x changes by one integer
now.

Press $\boxed{WINDOW}$ (see Fig. 4.21).

Calculate $X_{max} - X_{min}$. The distance is
exactly 94, the same as the number of pixel
jumps across the screen. This is a "friendly"
window. Refer back to Chapter 2 Section
2.2.4 for more on *friendly windows*.

Press $\boxed{ZOOM}$ $\boxed{4}$. This is the "decimal"

window. When you press $\boxed{TRACE}$ and move
the arrows, x now increments by .1, or one
decimal point. This is another "friendly"
window (see Fig. 4.22).

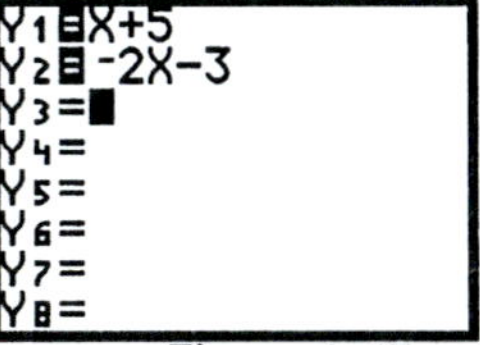

Figure 4. 17

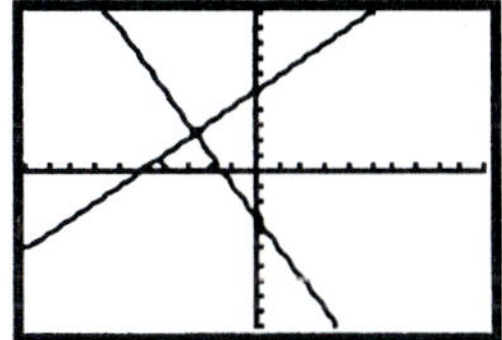

Figure 4. 18

Figure 4. 19

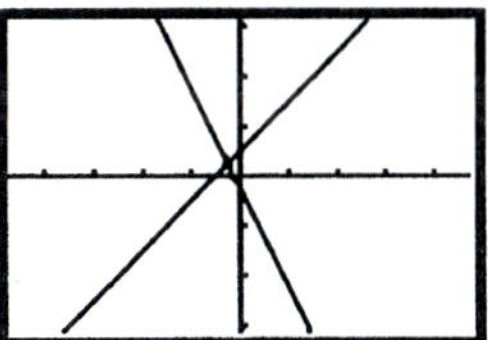

Figure 4. 20

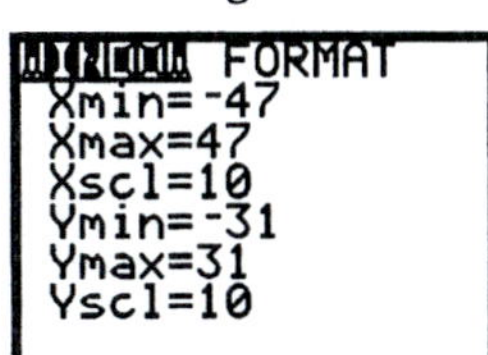

Figure 4. 21 Integer window

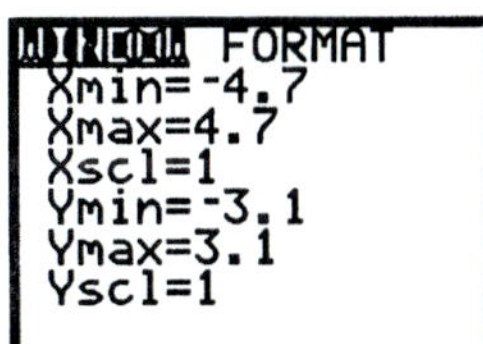

Figure 4. 22 Decimal window

4.2.2 Two Points Determine a Line

If two points (x_1, y_1) and (x_2, y_2) are known, you have enough information to determine the slope or average rate of change using the slope formula:

$$m = \text{slope} = \frac{(y_2) - (y_1)}{(x_2) - (x_1)}.$$

Once the slope is known, you can use $y = mx + b$ to find the equation of the line.

Example 3

If two points on the graph of a linear equation are (5, -1) and (-15, -13), determine the linear equation associated with these points.

1. Find the slope.

2. Substitute $m = 3/5$ into $y = mx + b$:
$$y = (3/5)x + b$$
3. Pick one of the points for values of x and y. The point (5, -1) means $x = 5$, and $y = -1$. Substitute these values into $y = mx + b$ and solve for b:
$$-1 = (3/5)\,(5) + b$$
$$-1 = 3 + b$$
$$b = -4$$
4. Write the linear function:
$$y = (3/5)\,x - 4$$

This is the linear function through the points (5, -1) and (15, 13) .

4.2.3 Use the Calculator to Verify the Equation of a Line Between Two Points with a Linear Regression Equation

CLEAR L1 and L2. Use a shortcut method:

Press STAT [1:Edit] Use △ to position the cursor on the word L1; press CLEAR ▽ ▷ .

Repeat for L2.

Example 4

Find the linear equation that goes through the points (5, -1) and (-15, -13).

1. Enter the points into L1 and L2. (see Fig. 4.23).
2. Find the linear regression equation.

Press STAT ▷ to <CALC> ; select

[:LinReg(ax+b)] (see Fig. 4.24). Press 2nd

L1 , 2nd L2 ENTER (see Figs. 4.25).

Note: The calculator uses "a" instead of "m". Both expressions represent a linear function: $\quad y = mx + b = ax + b$

In Figure 4.26 you see the values for the linear equation: $a = \text{slope} = m = .6$ and, $b = -4$ (the y - intercept).

Thus $y = 0.6x - 4$ is the same as the linear function you determined above:
$$y = (3/5)x - 4 = 0.6x - 4$$

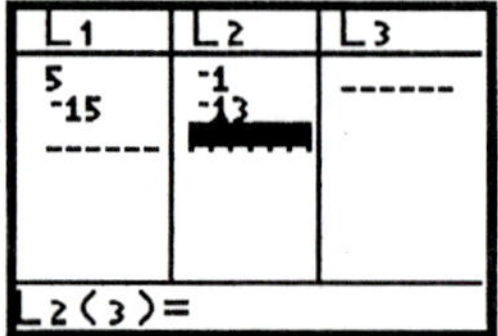

Figure 4. 23

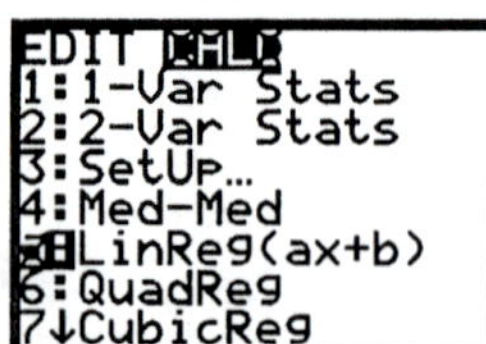

Figure 4. 24
On the TI-83 select [4:LinReg(ax+b)]

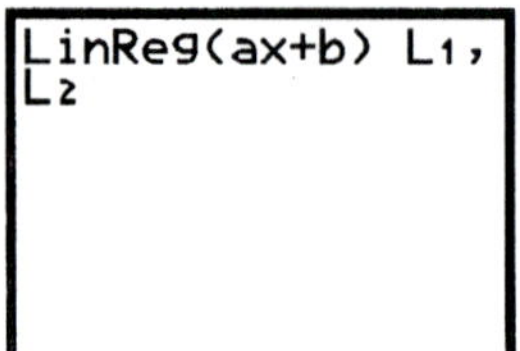

Figure 4. 25

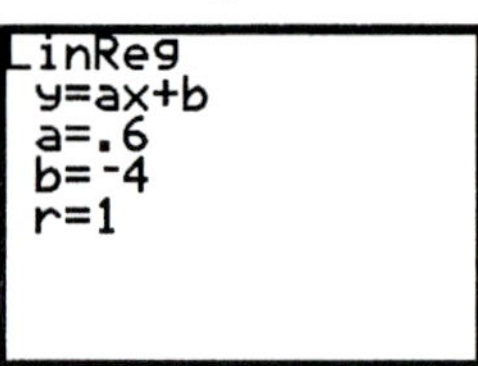

Figure 4. 26
The r value tells you how good of a fit you have. When $|r| = 1$ the fit is perfect, i.e. $r = 1$ or $r = -1$.

4.3 Special lines

4.3.1 Horizontal Lines

Horizontal lines are parallel to the x-axis and have a slope of 0. For this case, if $m = 0$ in the general equation $y = mx + b$, the equation becomes $y = 0x + b$ or $y = b$.

Example 5

Graph the following: Press $\boxed{Y=}$

$$Y1 = -3$$
$$Y2 = 3$$
$$Y3 = 2$$
$$Y4 = -1$$

(see Fig. 4.27). Press $\boxed{ZOOM}$; select [4:Zdecimal] (see Fig. 4.28).

Figure 4. 27

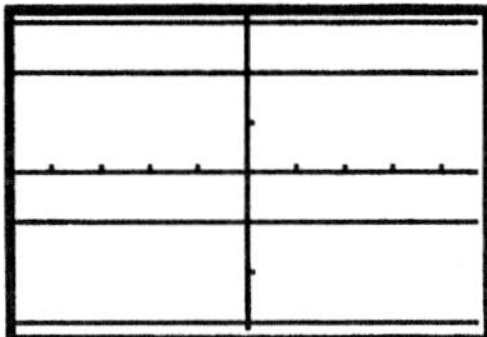

Figure 4. 28

4.3.2 Vertical Lines

Vertical lines are of the form $x = c$ or x is constant. y can be any number.

Example 6

Draw the vertical lines:
$$x = 3, x = -4 \text{ and } x = 1.$$

Since vertical lines are *not* functions the calculator must draw them, rather than graph them. The calculator can draw a vertical line using the DRAW menu.

CLEAR $\boxed{Y=}$. Press $\boxed{2nd}$ $\boxed{DRAW}$ select

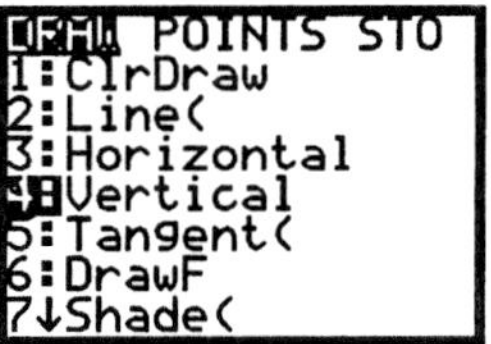

Figure 4. 29

[4:Vertical](see Fig. 4.29). Use $\boxed{\triangleright}$ to position the cursor on $x = 3$. Press $\boxed{ENTER}$.

Use $\boxed{\triangleleft}$ to reposition on $x = 1$. Press $\boxed{ENTER}$.

Use $\boxed{\triangleleft}$ to reposition on $x = -4$. Press $\boxed{ENTER}$.

You should see three vertical lines as in Figure 4.30.

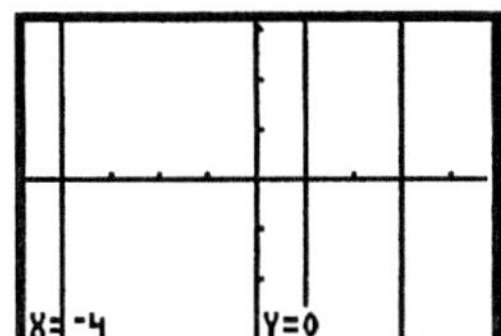

Figure 4. 30

4.3.3 Proportional Relationships

The graph of a line which passes through the origin (0, 0) has a y-intercept $= 0$. For this case, $b = 0$ or the equation is simply :

$$y = mx \text{ or } y/x = m$$

The ratio of y to x is constant (m) for every point on the graph line.

Example 7

Graph the following direct proportions:

$$Y1 = 3x$$
$$Y2 = (2/3)x$$
$$Y3 = -4\,x$$

All of the graphs in Figure 4.31 go through the origin. Their intercepts are zero. These are examples of proportional linear graphs.

4.3.4   Parallel lines

Parallel lines have the same slope.

Example 8

Graph the following:

$$Y1 = 3x$$
$$Y2 = 3x + 4$$
$$Y3 = 3x - 5$$
$$Y4 = 3x + 6$$

See Figure 4.32.

Press ZOOM ; select [6:ZStandard].

Look at the graphs of the lines (see Fig. 4.33). The lines are parallel with different y - intercepts.

Press TRACE then ∇ to see the intercepts.

4.3.5   Perpendicular Lines

If the graph of the equation, $y = (2/3)x + 5$ is rotated 90 degrees, the resulting line will have a slope of -3/2. In general, if the slope of a line is m_1, the slope of a perpendicular line is $m_2 = -1/m_1$ (the negative reciprocal of m_1).

> **Note:** For perpendicular lines, l_1 and l_2, the product of their slopes m_1 and m_2 will always equal -1. If $m_1 m_2 = -1$ then l_1 is perpendicular to l_2.

Example 9

Graph:

$$Y1 = (2/3)x + 5$$
$$Y2 = (-3/2)x + 5$$
$$Y3 = (-3/2)x - 4$$

Enter the above functions into Y= (see Fig. 4.34). Press GRAPH (see Fig. 4.35).

Due to the scaling, the graphs do not look perpendicular. To adjust the window so that the x and y scalings are proportional; press ZOOM ; select [5:ZSquare] as in Figure 4.36. Now the equations look perpendicular (see Fig. 4.37).

> **Note:** Both Y2 and Y3 are perpendicular to Y1. Y2 is parallel to Y3 because their slopes are equal.

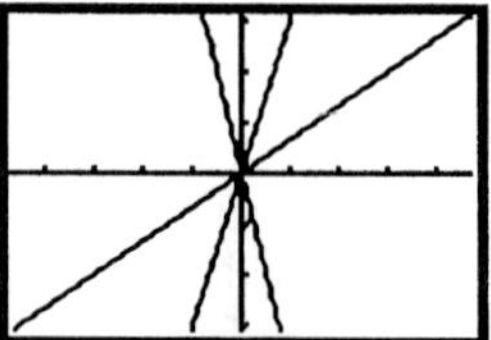

Figure 4. 31

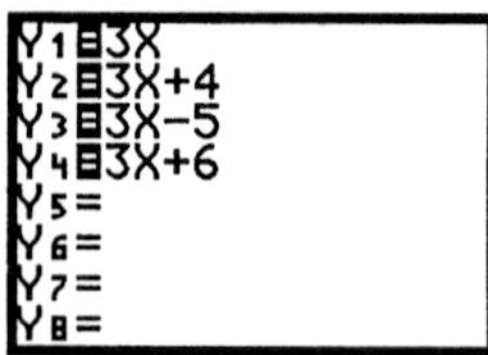

Figure 4. 32

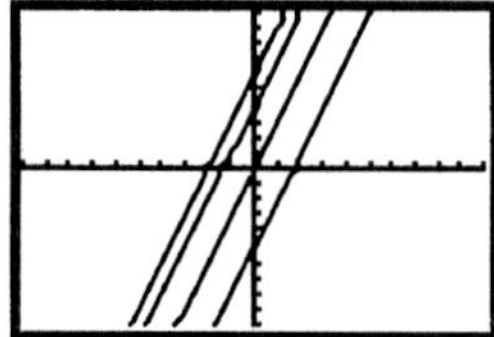

Figure 4. 33

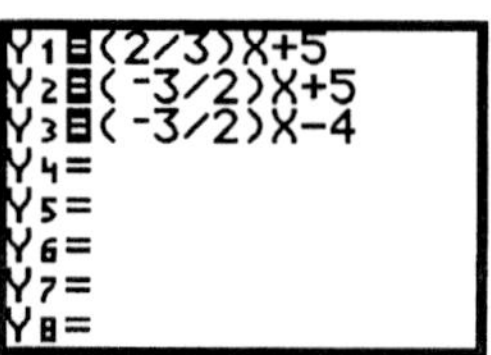

Figure 4. 34

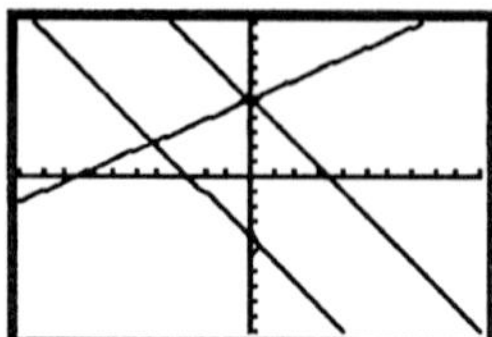

Figure 4. 35

Figure 4. 36

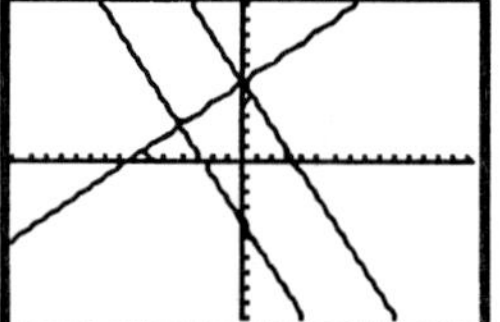

Figure 4. 37

4.3.6 Entering Linear Equations Not in Function Form

The equation $y = mx + b$ is called the slope-intercept form of a linear equation, or the function form, since the equation is solved for y. A linear relationship may also be described by an equation in the *standard form*:

$$Ax + By = C.$$

Example 10

Transform the equation $-2x + 3y = 15$ into function form so that you can enter it into your calculator and graph it.

$$-2x + 3y = 15$$

$3y = 2x + 15$	add $2x$
$y = (2/3)x + 15/3$	divide by 3
$y = (2/3)x + 5$	simplify

$-2x + 3y = 15$ is equivalent to $y = (2/3)x + 5$ which you graphed above.
Refer back to Figures 4.34 to see how the function was entered. See Figure 4.38 for the graph of $y = (2/3)x + 5$ by itself.

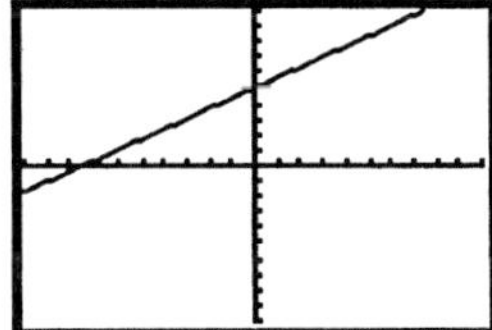

Figure 4. 38
The graph of Y1 = $(2/3)x + 5$

4.3.5 Fractional Coefficients of Linear Functions

Notice that in all of the examples above when the coefficient of x was a fraction, you enclosed the fraction in parentheses. See what happens if you do not use parentheses. The TI-82 and TI-83 behave differently. Enter the function $y = 1/2x$.

Press ⎡ZOOM⎤ ⎡4⎤ .

The TI-82 interprets this as $y=1/(2x)$ (see Fig. 4.39); while the TI-83 interprets the function as $y=(1/2)(x)$ as in Figure 4.40. **As a precaution always put fractions in parentheses**.

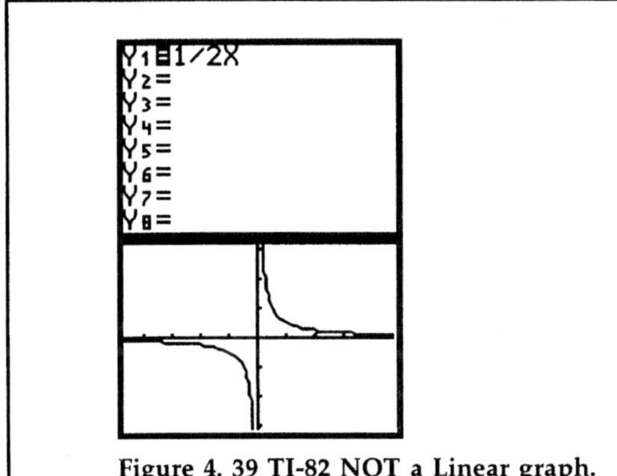

Figure 4. 39 TI-82 NOT a Linear graph.

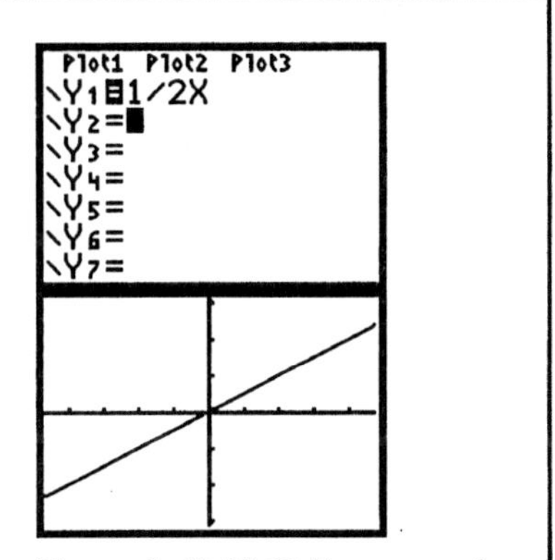

Figure 4. 40 TI-83 linear graph.

Chapter 5
Linear Regression Equations

5.1 Linking Calculators

This course comes with programs that store data to the graphing calculator. Once the data is stored to lists, you can create plots and find regression equations on the data. Your instructor will be able to download the programs from a computer to a calculator. Your calculator came with a cable that allows you to link calculators. Once linked, you can receive and send programs as well as data lists.

5.1.1 Receiving Data

1. Attach the cable to both calculators. Be sure to push the cable **all** the way in.

2. Press [2nd] [LINK] [▷] to <RECEIVE>. Press [ENTER] (see Fig. 5.1).

3. The receiving calculator must say *Waiting...*

5.1.2 Sending data:

1. Press [2nd] [LINK] ; select [3:Select Current...].

2. [∇] to the programs or lists to be sent. Press [ENTER] to select. A small square indicates the selection has been made (see Fig. 5.2).

3. [▷] to <TRANSMIT> press [ENTER] (see Fig. 5.3).

4. Wait for the message *Done..* on the receiving calculator (see Fig. 5.4).

5.1.3 Running a Program

Press [PRGM] , then select the program.

Press [ENTER] (see Figs. 5.5 and 5.6).

The program has stored the FAM1000 data to List1 though List6.

Press [STAT] select [1:Edit] to see the data in the lists (see Fig. 5.7).

Figure 5. 1

Figure 5. 2

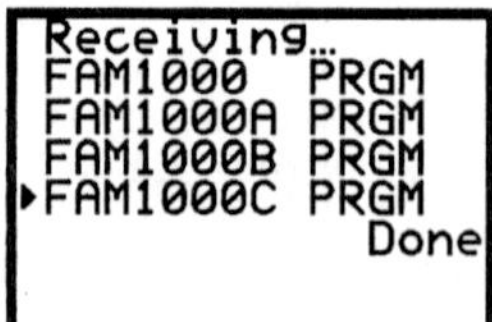

Figure 5. 3

Figure 5. 4

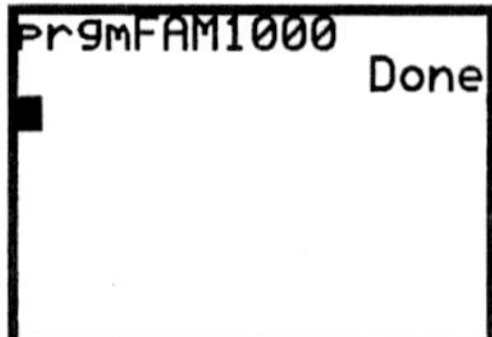

Figure 5. 5

Figure 5. 6

To see the program press $\boxed{\text{PRGM}}$ $\boxed{\triangleright}$ to
<EDIT>. Select the program to view. Press
$\boxed{\text{ENTER}}$ (see Fig. 5.8).

The program will look like Figure 5.9. At
the end of this chapter the complete FAM
1000 data sets will be listed with
appropriate column headings.

> **Trouble shooting:** If you get lost and don't
> know which menu you are in, press $\boxed{\text{2nd}}$
> $\boxed{\text{QUIT}}$ and start over.

5.2   Linear Regression Equations

The FAM 1000A program stores the years of
education in L1 and the mean personal
income in L2. To plot the data press $\boxed{\text{2nd}}$
$\boxed{\text{STATPLOT}}$ $\boxed{\text{ENTER}}$ (see Fig. 5.10).

5.2.1   Set up the plot
1.   Select **ON** $\boxed{\text{ENTER}}$.
2.   Select Type: **scatterplot icon,** $\boxed{\text{ENTER}}$.
3.   Select Xlist: **L1** $\boxed{\text{ENTER}}$.
4.   Select Ylist: **L2** $\boxed{\text{ENTER}}$.
5.   Select Mark: □   (see Fig. 5.11).

> **Trouble shooting:**
> CLEAR $\boxed{\text{Y =}}$ and turn **off** all other plots.

5.2.2   Draw a Scatter Plot
Press $\boxed{\text{ZOOM}}$; select [9:ZoomStat] (see Fig.
5.12 and 5.13). Review Chapter 2 Section
2.2.5 for selecting appropriate viewing
windows.

5.2.3   Find the Equation
Press $\boxed{\text{STAT}}$ $\boxed{\triangleright}$ to <CALC>, select
[:LinReg(ax+b)] $\boxed{\text{2nd}}$ $\boxed{\text{L1}}$ $\boxed{,}$ $\boxed{\text{2nd}}$ $\boxed{\text{L2}}$
$\boxed{\text{ENTER}}$ (see Figs 5.14 through 5.16).

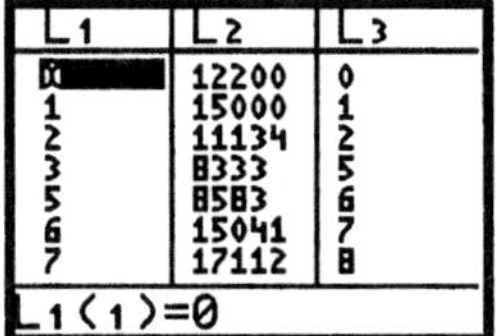

Figure 5. 7

Figure 5. 8

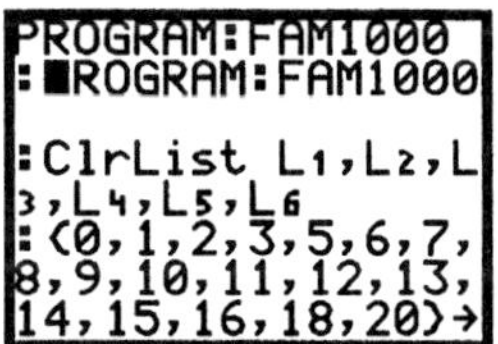

Figure 5. 9

Figure 5. 10

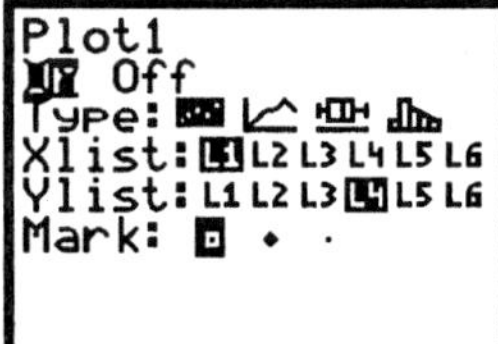

Figure 5. 11

Figure 5. 12

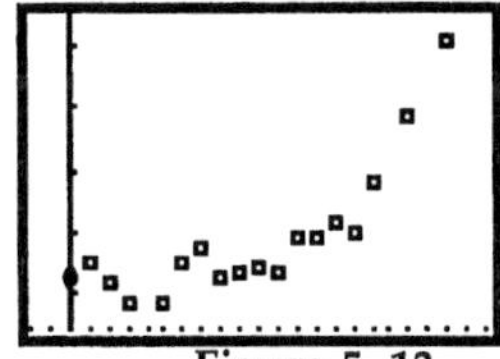

Figure 5. 13

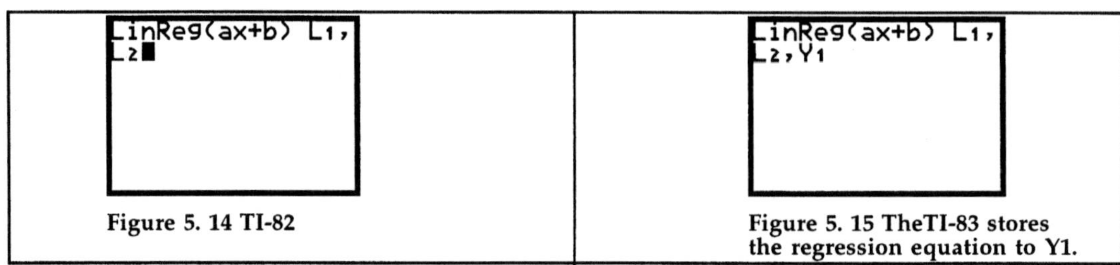

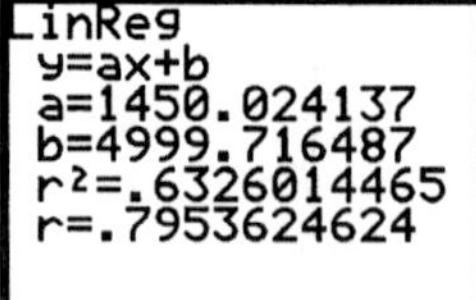

Figure 5. 14 TI-82

Figure 5. 15 TheTI-83 stores the regression equation to Y1.

Enter the equation into Y1. For the TI-82 press $\boxed{Y=}$ $\boxed{VARS}$; select [5:statistics] $\boxed{\triangleright}$ $\boxed{\triangleright}$ to <EQ> ; select [:RegEQ] (see Fig. 5.17).

Recall $a = m$, the slope or rate of change and $b =$ the y-intercept.

Press $\boxed{GRAPH}$.

Figure 5.18 shows both the scatterplot and the graph of the regression equation.

The equation means that on average for every year of education a person earns an additional $1450.02 and with no education a person earns only $4999.72 . This corresponds to the slope of the equation, which is 1450.02, and the y-intercept which is 4999.72.

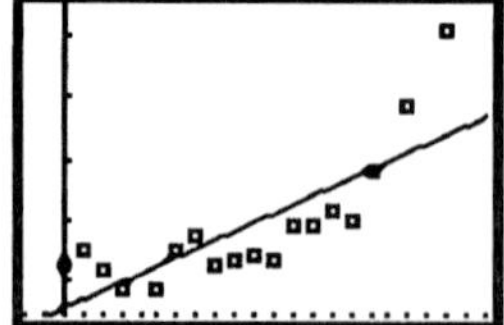

Figure 5. 16 TI-83 screen

The TI-83 linear regression equation. The r value tells how good of a fit you have. To see the r values: press $\boxed{2nd}$ $\boxed{CATALOGUE}$ choose DiagnosticsOn.

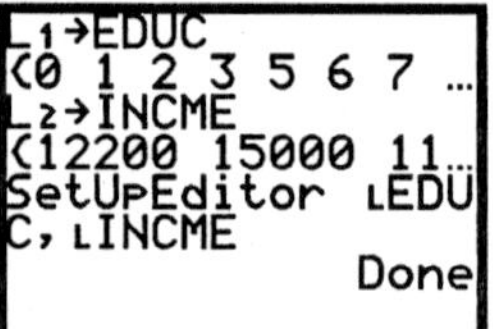

Figure 5. 17

Figure 5. 18

5.3   TI-83 Naming Lists

The TI-83 allows you to name lists and store them. Press $\boxed{2nd}$ $\boxed{L1}$ $\boxed{STO\triangleright}$ $\boxed{2nd}$ $\boxed{ALPHA}$ then type no more that 5 letters. Repeat for list 2 , L2 (see Fig. 5.19). To place the list in the edit screen press $\boxed{STAT}$; select [5:SetUpEditor], $\boxed{2nd}$ $\boxed{LIST}$ then choose the names, $\boxed{ENTER}$ (see Fig. 5.19).Press $\boxed{STAT}$ [1:Edit]to see the named lists (see Fig. 5.20).

Figure 5. 19

Figure 5. 20

This data is also on the CD-ROM in the back of your text.

FAM 1000A: Mean Personal Wages					
Years of Education	Mean Personal Wages ($) (all)	Years of Education	Mean Personal Wages ($) (men)	Years of Education	Mean Personal Wages ($) (women)
L1	L2	L3	L4	L5	L6
1	$10,811	1	$10,811	4	$10,400
4	$10,600	4	$11,000	6	$6,846
6	$12,095	6	$15,011	8	$6,395
8	$5,664	8	$5,225	9	$5,652
9	$6,817	9	$8,111	10	$8,941
10	$12,883	10	$17,876	11	$6,471
11	$9,834	11	$11,596	12	$16,154
12	$20,001	12	$23,848	14	$18,736
14	$23,031	14	$27,103	16	$21,224
16	$33,840	16	$46,607	18	$27,058
18	$36,913	18	$49,042	20	$25,500
20	$45,635	20	$49,137		

FAM 1000B: Mean Personal Total Income					
Years of Education	Mean Personal Total Income ($) (all)	Years of Education	Mean Personal Total Income ($) (men)	Years of Education	Mean Personal Total Income ($) (women)
L1	L2	L3	L4	L5	L6
1	$10,811	1	$10,811	4	$10,400
4	$10,600	4	$11,000	6	$7,129
6	$12,196	6	$15,011	8	$6,711
8	$11,148	8	$13,811	9	$7,536
9	$16,550	9	$26,566	10	$10,433
10	$14,402	10	$19,430	11	$7,463
11	$13,605	11	$16,823	12	$17,688
12	$22,773	12	$27,859	14	$21,760
14	$26,103	14	$30,220	16	$24,986
16	$40,009	16	$55,211	18	$34,649
18	$44,212	18	$55,981	20	$26,025
20	$59,228	20	$65,003		

FAM 1000C: Mean Personal Total Income for White and Nonwhite Men					
Years of Education	Mean Personal Total Income ($) All Men	Years of Education	Mean Personal Total Income ($) All White Men	Years of Education	Mean Personal Total Income ($) All Nonwhite Men
L1	L2	L3	L4	L5	L6
1	$10,811	1	$10,811	8	$11,880
4	$11,000	4	$11,000	9	$19,256
6	$15,011	6	$15,011	10	$25,277
8	$13,811	8	$14,025	11	$8,727
9	$26,566	9	$28,655	12	$35,808
10	$19,430	10	$19,012	14	$29,007
11	$16,823	11	$18,172	16	$29,809
12	$27,859	12	$26,872	18	$19,530
14	$30,220	14	$30,351	20	$40,250
16	$55,211	16	$56,152		
18	$55,981	18	$60,736		
20	$65,003	20	$67,360		

FAM 1000D: Mean Personal Total Income for White and Nonwhite Women					
Years of Education	Mean Personal Total Income ($) All Women	Years of Education	Mean Personal Total Income ($) All White Women	Years of Education	Mean Personal Total Income ($) All Nonwhite Women
L1	L2	L3	L4	L5	L6
4	$10,400	4	$10,400	8	$11,269
6	$7,129	6	$7,129	9	$8,490
8	$6,711	8	$5,800	10	$13,001
9	$7,536	9	$7,128	11	$19,061
10	$10,433	10	$9,952	12	$14,613
11	$7,463	11	$4,886	14	$20,918
12	$17,688	12	$18,325	16	$17,295
14	$21,760	14	$21,926	18	$28,573
16	$24,986	16	$26,129		
18	$34,649	18	$35,775		
20	$26,025	20	$26,025		

Chapter 6
Intersecting Lines

6.1   Intersecting Lines

Lines that cross each other are said to *intersect*. All lines eventually intersect unless they are parallel lines or they are the same line.

Example 1

Graph the following two functions and determine the point of intersection, if it exists.

$$f(x) = 3x + 4 \quad \text{and} \quad g(x) = 0.5x - 1$$

Press $\boxed{Y=}$ $\boxed{CLEAR}$ to clear all the old functions. Enter the above equations into Y1 and Y2 (see Figure 6.1).

Press $\boxed{ZOOM}$; select [6: ZStandard] . Press $\boxed{TRACE}$ then use $\boxed{\nabla}$ or $\boxed{\Delta}$ to estimate the point of intersection. It looks like the graphs cross when x is about -2 and y is about -2 (see Fig. 6.2).

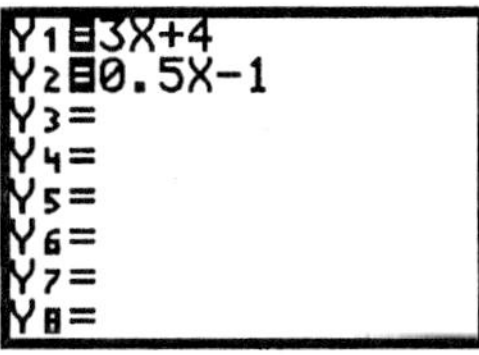

Figure 6. 1

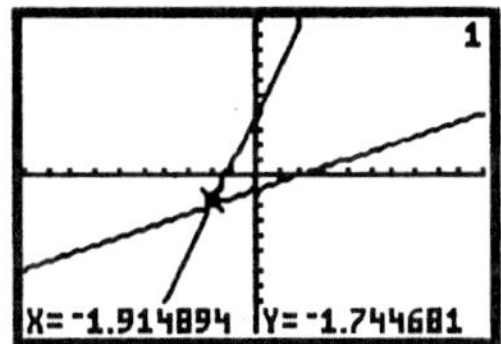

Figure 6. 2

6.1.1   The Calculate Menu

Use the calculator to find the point of intersection.

Press $\boxed{2nd}$ $\boxed{CALC}$, select [5:intersect] (see Fig. 6.3).

The calculator prompts you:

1. Select the first curve, press $\boxed{ENTER}$.

2. Select the second curve, press $\boxed{ENTER}$.

3. Using the arrows move the cursor near the point of intersection (your guess).

     Press $\boxed{ENTER}$ (see Fig. 6.4 and 6.5).

The point of intersection is (-2,-2).

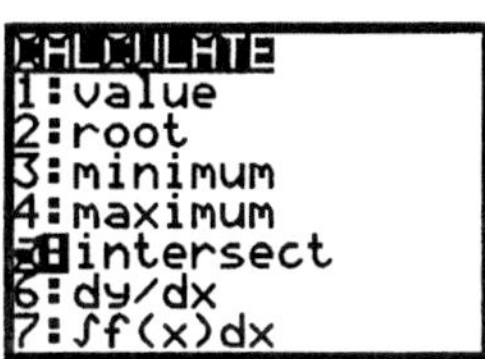

Figure 6. 3

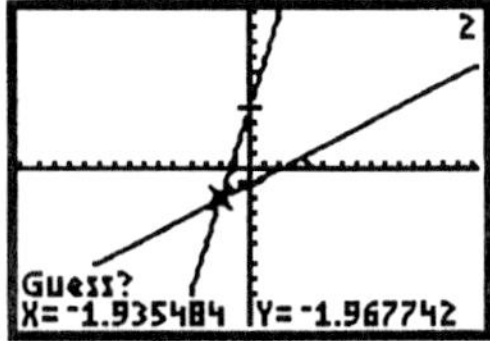

Figure 6. 4

6.1.2   Algebraic Verification

Check algebraically:

If $x = -2$ then

Y1= $3x + 4$ becomes   Y1 = 3(-2) + 4 = -2
Y2= $0.5x - 1$ becomes   Y2 = 0.5(-2) - 1 = -2

Both equations have the same y values so the solution is $x = -2$ and $y = -2$ or the point of intersection is (-2,-2).

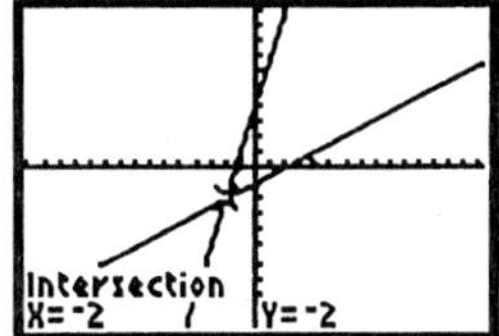

Figure 6. 5

6.1.3 Solving Algebraically

Since Y1=Y2 at the point of intersection, you solve algebraically by substitution:

$$3x + 4 = 0.5\,x - 1$$
$$2.5\,x + 4 = -1 \qquad \text{add } -0.5\,x$$
$$2.5\,x = -5 \qquad \text{add } -4$$
$$x = -5/2.5 = -2 \qquad \text{divide by 2.5}$$

Substitute $x = -2$ in Y1 and Y2 as above to find $y = -2$. The solution is (-2, -2).

6.1.4 Adjusting the Window

Example 2

Find the point of intersection for the system of equations :

$$h(x) = 0.025x - 25 \quad \text{and}$$
$$k(x) = -2x + 50$$

Enter the expressions into Y1 and Y2. Press GRAPH (see Figures 6.6 and 6.7).

The graphs do not appear! Where are they? Press TRACE . This will give you points on the graph (see Fig. 6.8). Using algebra, let $x = 0$ to find the y - intercepts for the graphs at (0, -25) and (0, 50). To see these points we must choose Ymin and Ymax beyond those points. Set the WINDOW to Figure 6.9. Now we see parts of the graphs. The point of intersection appears to be further to the right (see Fig. 6.10).

Press 2nd TABLE for more information. Y1 is climbing very slowly ($m = 0.025$) while Y2 is decreasing and somewhere around $x = 37$ they are about equal (see Fig. 6.11). Experiment with different window settings. Set Xmax so that the point of intersection and beyond can be seen. Press GRAPH and repeat the CALC Menu steps in Section 6.1.1 (see Fig. 6.12).

6.1.5 Solve Algebraically to Confirm

Let Y1=Y2:

$$0.025x - 25 = -2x + 50$$
$$2.025x - 25 = 50$$
$$2.025x = 75$$
$$x = 75/2.025 = 37.0370370\ldots$$

Substitute the x value into Y1 and Y2. To find the value of y.

Y1= 0.025(75/2.025) - 25 = -24.07407407...
Y2= -2(75/2.025) + 50 = -24.07407407...

Although the algebra is pretty fast the arithmetic could still use some help from a calculator!

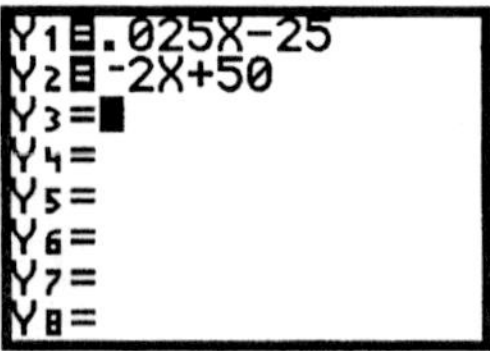

Figure 6. 6

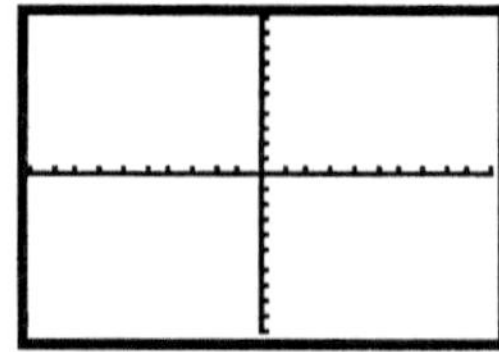

Figure 6. 7

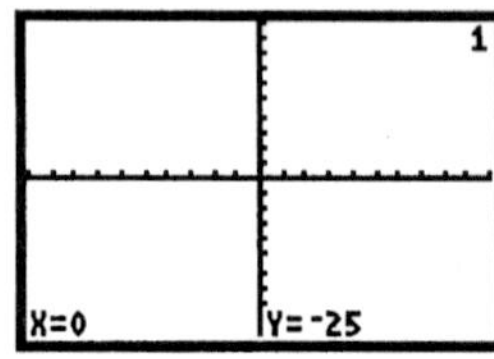

Figure 6. 8

WINDOW FORMAT
Xmin=-10
Xmax=10
Xscl=1
Ymin=-60
Ymax=60
Yscl=10

Figure 6. 9

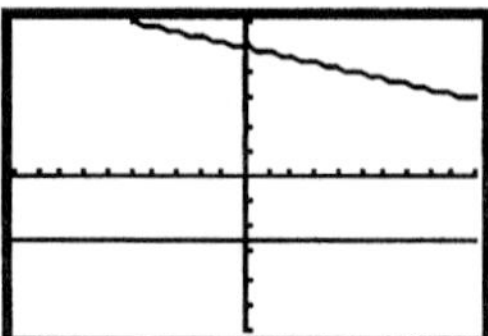

Figure 6. 10

X	Y1	Y2
33	-24.18	-16
34	-24.15	-18
35	-24.13	-20
36	-24.1	-22
37	-24.08	-24
38	-24.05	-26
39	-24.03	-28

X=37

Figure 6. 11

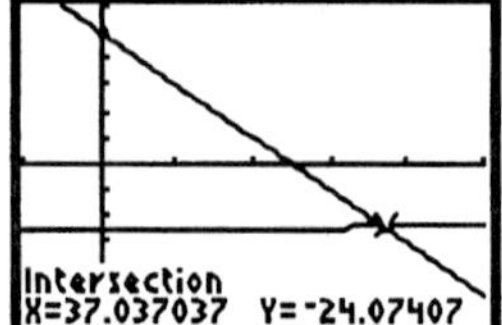

Figure 6. 12

6.2 Piecewise Functions

The graphing calculator can graph piecewise functions very easily. However, it is important to understand how the graphing calculator performs a test.

6.2.1 The TEST Menu

The graphing calculator can tell if a statement is true or false using the TEST menu. Notice that you find the equal and inequality symbols here. Press $\boxed{\text{2nd}}$ $\boxed{\text{TEST}}$ (see Fig. 6.13).

Go to the Home Screen. Press $\boxed{\text{2nd}}$ $\boxed{\text{QUIT}}$.

Example 3
Determine if the following are True or False by typing the following:

      a. $5 = 5$
      b. $5 \leq 7$
      c. $5 = 4$
      d. $5 \geq 7$

See Figures 6.14 and 6.15.

Note: The calculator is performing a test. It tells you **1** for **True** and **0** for **False**.

This method can be used to enter a piecewise function (a function defined in pieces) into the calculator.

Example 4
Graph the piecewise function:

$$f(x) = \ y = \begin{cases} x + 2 \text{ for } x > 0 \\ -x - 3 \text{ for } x \leq 0 \end{cases}$$

y is defined under two conditions:

    1.   when $x > 0$, use $y = x + 2$
    2.   when $x \leq 0$, use $y = -x - 3$

Press $\boxed{\text{Y=}}$ and enter the equations as in Figure 6.16 .

Since this is a piecewise function, if $x > 0$ choose the function $y1 = x + 2$ but if $x \leq 0$ choose the function $y2 = -x - 3$ (see Fig. 6.17).

Note: On the calculator we will use the division sign to enter the x value condition. The calculator will perform a test by putting 1 or 0 in the denominator. When the denominator is 0 the function is undefined and no points will be drawn. When the denominator is 1 the function is plotted.

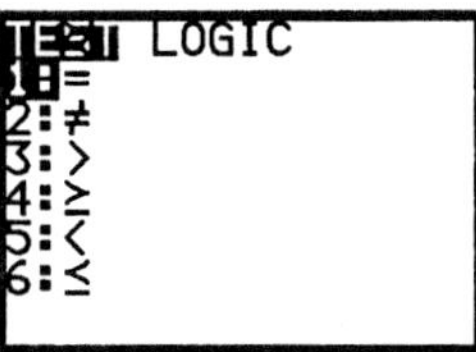

Figure 6. 13

Figure 6. 14

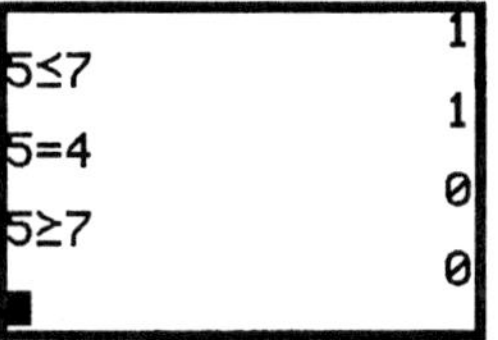

Figure 6. 15

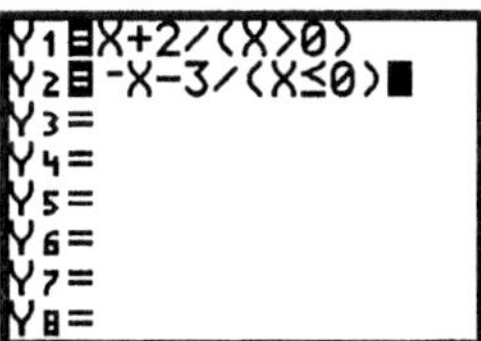

Figure 6. 16

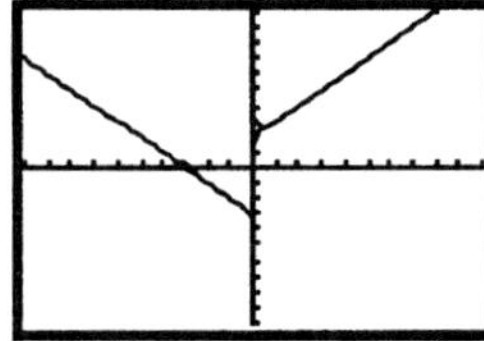

Figure 6. 17

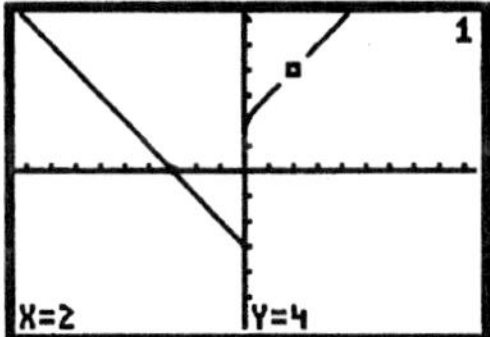

Figure 6. 18

The graph of *f(x)* is in two pieces.
When you are on Y1 you are on the graph of
condition one, or *f(x) = x + 2.*

Press $\boxed{\text{TRACE}}$ (see Fig. 6.18).

When you are on Y2 you are on the graph of
condition two, or *f(x) = -x - 3.*

Press $\boxed{\nabla}$, then $\boxed{\text{TRACE}}$ $\boxed{\triangleleft}$, to duplicate
Figure 6.19. Notice that when the
condition no longer applies, no *y* value is
given. For example if *x* = 3 condition two no
longer applies (see Fig. 6.20).
To find the value of *y* for *x* = 3, you must

switch to condition one. $\boxed{\triangle}$ to the graph of
Y1 (see Fig. 6.21).

6.2.2   An Alternate Method for Graphing Piecewise Functions

Many people choose to write a piecewise
function all on the same line since *f(x)* is
defined for all *x*. This entails making an

adjustment to the mode. Press $\boxed{\text{MODE}}$;

select **Dot** $\boxed{\text{ENTER}}$ (see Fig. 6.22).

Example 5
Graph the piecewise function

$$f(x) = \begin{cases} (x+3)^2 - 5 & \text{for } x < -2 \\ 2x + 6 & \text{for } x \geq -2 \end{cases}$$

Press $\boxed{\text{Y=}}$ $\boxed{\text{CLEAR}}$ to clear all expressions.

Type the piecewise function commands as in

Figure 6.23. Press $\boxed{\text{GRAPH}}$ (see Fig. 6.24).

The advantage of this method is that you
can TRACE on the function in the normal
manner.

Troubleshooting:
1. If you are in <u>connected</u> MODE the
    calculator tries to connect the point
    from the end of one piece of the graph
    to the end of the other piece, giving the
    false impression that the function is
    continuous (see Fig. 6.25).
2. Since you use <u>multiplication</u> for the
    test, the calculator places a 1 or a 0
    inside the test parentheses. This
    sometimes gives the false value of 0
    rather than an undefined value for the
    function.

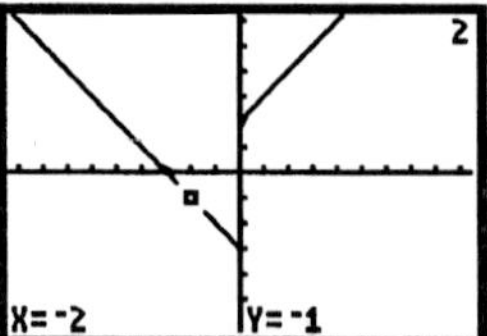

Figure 6. 19

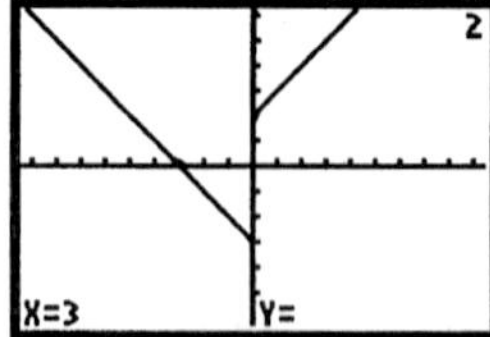

Figure 6. 20

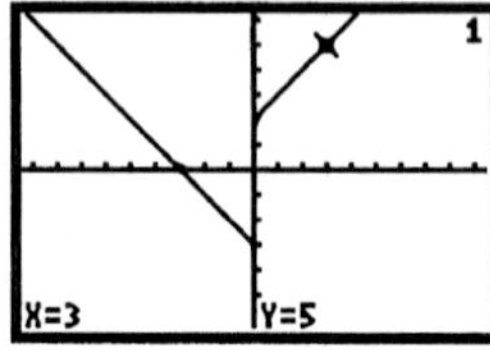

Figure 6. 21

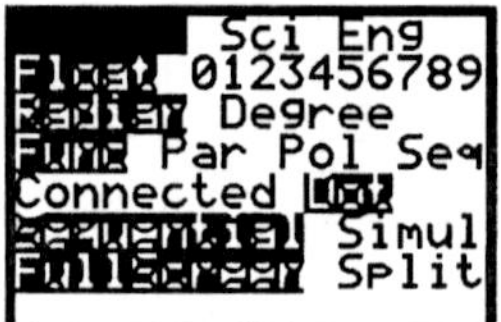

Figure 6. 22

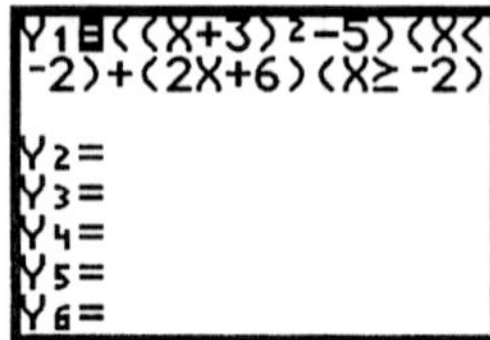

Figure 6. 23 See Troublesooting.

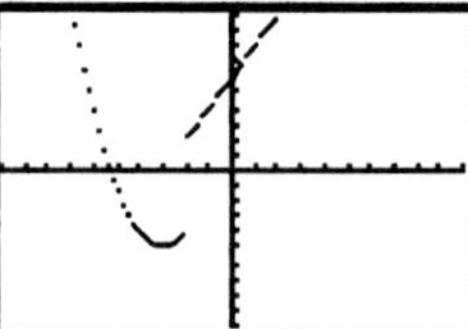

Figure 6. 24 Graph in Dot mode.

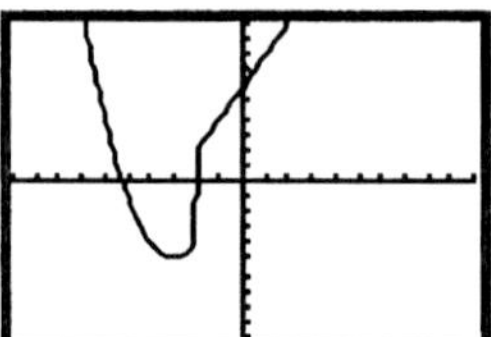

Figure 6. 25
False Graph in <u>Connected</u> mode.

Example 6

An 8% flat income tax is represented by $f(x)$. Under the flat tax everyone pays 8%, regardless of how much money is earned per year. A graduated income tax is represented by the piecewise function $g(x)$. Under this plan the first $20,000 is tax free, then between $20,000 and $100,000, 5% tax is paid. If you make beyond $100,000 you pay a 10% tax. Graph these functions.

$$f(x) = .08x \quad \text{for } x \geq 0$$

$$g(x) = \begin{cases} 0 & \text{for } 0 \leq x \leq 20000 \\ .05x(x - 20000) & \text{for} \\ & 20000 < x \leq 100000 \\ 4000 + .10(x - 100000) & \text{for} \\ & x > 100000 \end{cases}$$

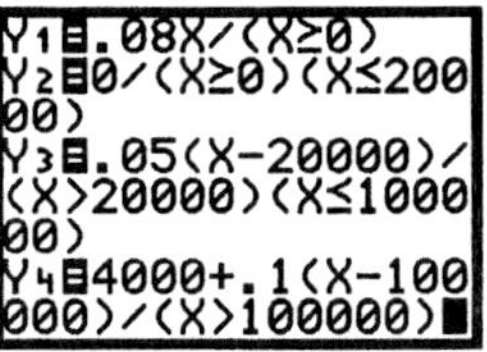

Figure 6. 26

Enter the piecewise functions into the graphing calculator as individual functions with separate conditions.

Enter the function as shown in Figure 6.26.

$Y1 = 0.08x / (\ x \geq 0)$

$Y2 = \quad 0/(x \geq 0)(x \leq 20000)$

$Y3 = 0.05(x - 20000) / (x > 20000)(x \leq 100000)$

$Y4 = 4000 + .1(x - 100000) / (x > 100000)$

6.2.3   Adjust the Viewing Window

Look at the values for x. They go beyond $100,000. Evaluate Y1 and Y4 for $x = 110,000$ (see Fig. 6.27). With this new information about x and y set the viewing window.

Press WINDOW ∇ ; set as in Figure 6.28.

Press GRAPH (see Fig. 6.29).

6.2.4   Find the Intersection Point

To see the point of intersection change your WINDOW to Xmax = 350000 and Ymax = 30000. Press GRAPH (see Fig. 6.30).

Use 2nd CALC [5:intersect] to find the point of intersection. (Refer back to section 6.1.1 if needed). You will need to select Y1 and Y4 as the pieces of the graphs that intersect (see Fig. 6.30) for the coordinates of the point. This means that under the flat tax system people earning less than $300,000 per year pay more taxes than under the graduated tax plan. The graduated tax is a better system for this group of tax payers.

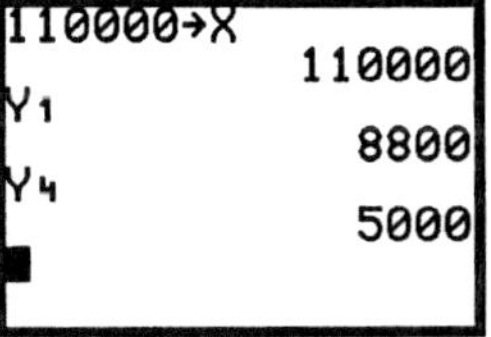

Figure 6. 27

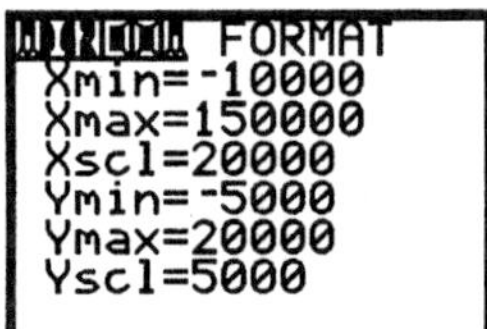

Figure 6. 28

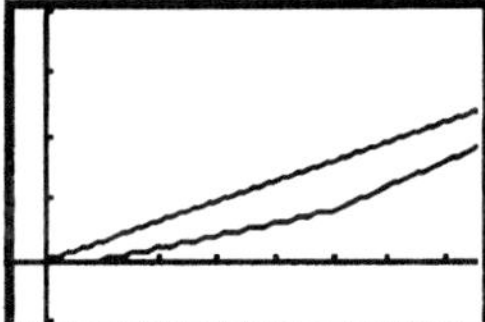

Figure 6. 29

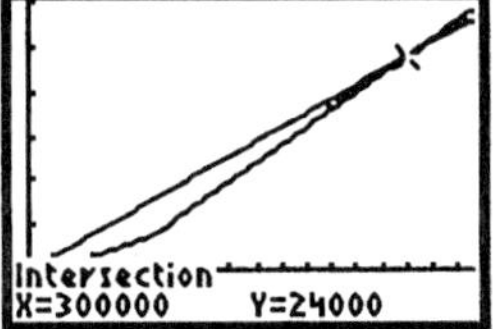

Figure 6. 30

Chapter 7
Scientific Notation and Exponent Properties

7.1   Scientific Notation

Scientific notation expresses large numbers and small numbers using powers of ten : $3250000000 = 3.25 \times 10^9$ and $0.00000123586 = 1.23586 \times 10^{-6}$. This can be done on the calculator using either $\boxed{10^x}$, the power of ten key or $\boxed{EE}$, the exponent of ten key.

> **Note**: Any decimal number can be written in scientific notation using the form $K \times 10^x$, where $1 \leq K < 10$ and x is an integer. To change back to decimal number form: if $x > 0$ move x decimal places to the right, if $x < 0$ move x decimal places to the left.

Example 1:
Type the following numbers:
1. 3.25×10^9
2.   0.00000123589

Press 3.25 $\boxed{2nd}$ $\boxed{10^x}$ 9 $\boxed{ENTER}$, then press 3.25 $\boxed{2nd}$ $\boxed{EE}$ 9 $\boxed{ENTER}$. Compare the results (see Fig. 7.1 or 7.2). When .00000123589 is typed, it is changed to scientific notation.

> **Troubleshooting:**
> When $x \geq 10$ or $x \leq -4$, in 10^x, the number is written in scientific notation.

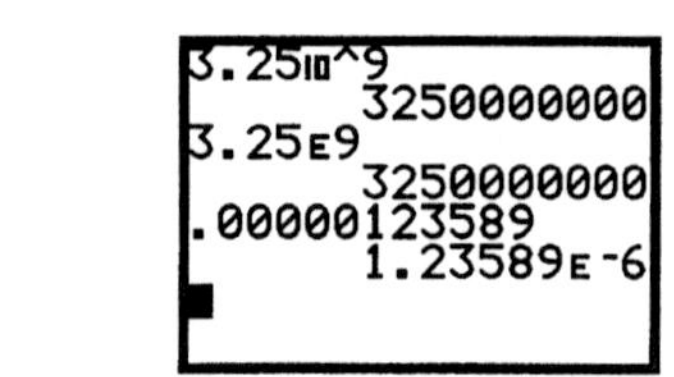

Figure 7.1 TI-82

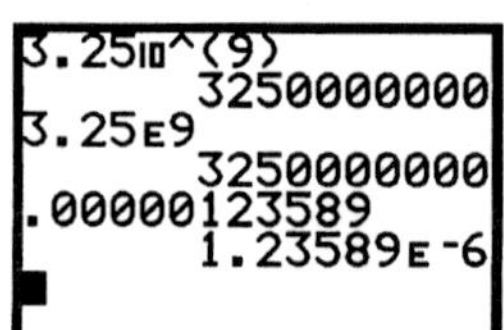

Figure 7. 2
The TI-83 will automatically insert a parenthesis when an exponent is used.

7.2   Verifying Properties of Exponents

Example 2
Verify numerically that the following are true by choosing values for a.
1. $a^0 = 1$
2. $a^{-1} = 1/a$
3. $a^6 = a \bullet a \bullet a \bullet a \bullet a \bullet a$

Let a assume various values. Enter the problems as in Figures 7.3 and 7.4. For the exponent use the $\boxed{\wedge}$ key.

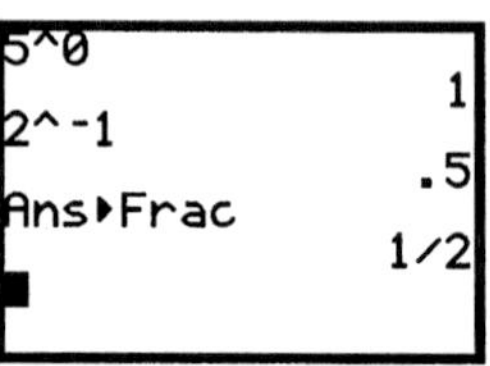

Figure 7. 3

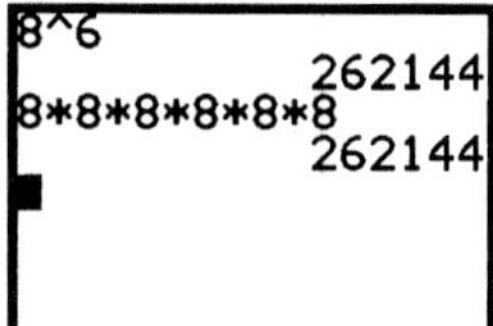

Figure 7. 4

7.2.1 Other Exponent Keys

There are other shortcut keys for exponents. The x^2 and x^{-1} keys are used to paste the exponents without using the $\wedge$ key (see Fig. 7.5). The cubic power and radical symbol are found under MATH (see Fig. 7.6).

Figure 7. 5

7.2.2 Fractional Exponents
Example 3
Show that the following are equivalent
1. $25^{1/2} = \sqrt{25}$
2. $32^{1/5} = \sqrt[5]{32}$

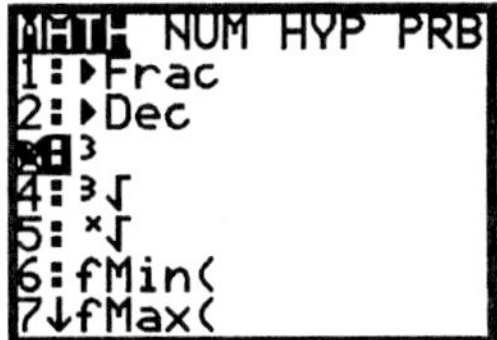

Figure 7. 6

Trouble Shooting: Fractional exponents **must** be enclosed in parentheses.

To access the square root symbol for $\sqrt{25}$, press 2nd √ 25 ENTER (see Fig. 7.7).

Roots other than square root are found under the MATH menu. To type $\sqrt[5]{32}$, press 5 MATH ; select [5: $\sqrt[x]{}$] 32 ENTER (see Fig. 7.8).

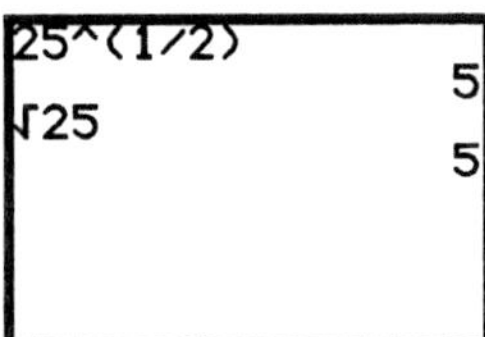

Figure 7. 7

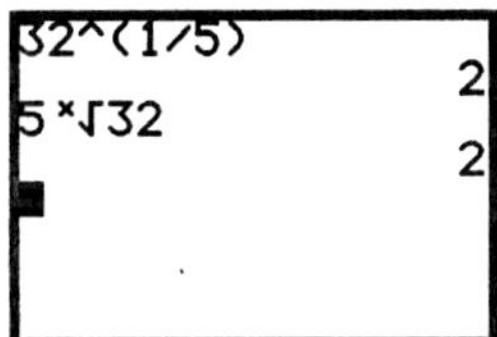

Figure 7. 8

Trouble Shooting:
The TI-82 has problems raising a negative base to a fractional power other than $1/n$. For $a^{m/n}$ where $m \neq 1$, the domain is restricted to $a \geq 0$. If $a < 0$ you get an **ERR:DOMAIN** message. You have to trick the calculator into performing the operation. The TI-83 does not have this problem. For the procedure to raise -8 to the 2/3 power, see Figures 7.9 and 7.10.

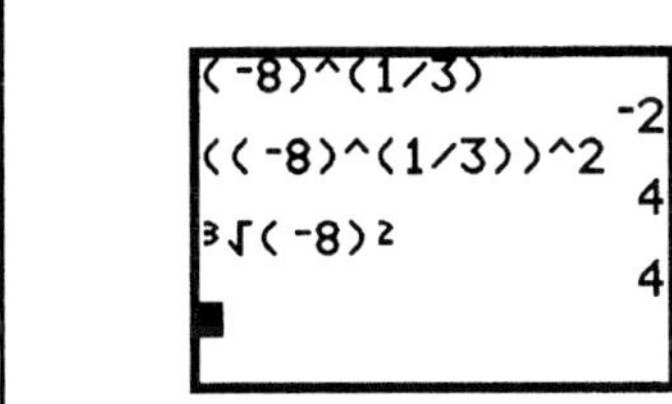

Figure 7. 9 TI-82

The TI-82 does NOT allow (-8)^(2/3), rewrite as a power to a power.

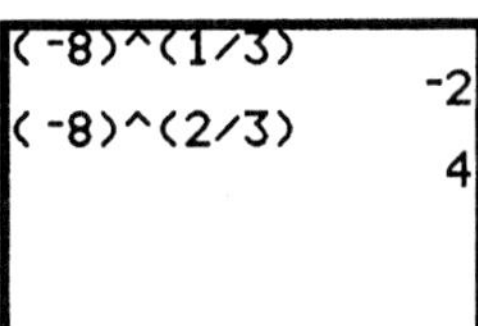

Figure 7. 10 TI-83

The TI-83 accepts (-8)^(2/3).

7.3   Using the Logarithm Key.

To find the exponent or power of ten in an equation, we use logarithms to "undo" the exponent.

If $10^x = N$ then $\log_{10} N = x$

Example 4

Write $10^x = 25$ as a logarithmic equation.

A logarithm is the value of the exponent. The solution is: $\log_{10} 25 = x$.
You read the above equation as "The logarithm of 25 to base 10 is x". To find the value of the exponent press $\boxed{\text{LOG}}$ 25 (see Fig. 7.11).

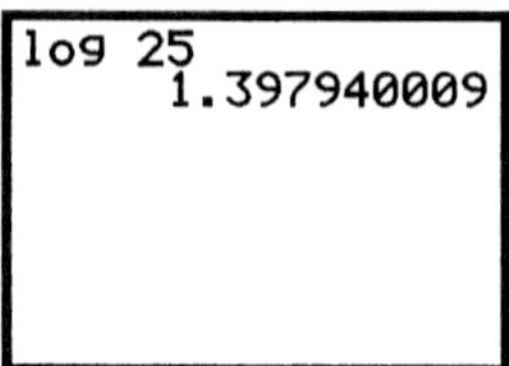

Figure 7. 11

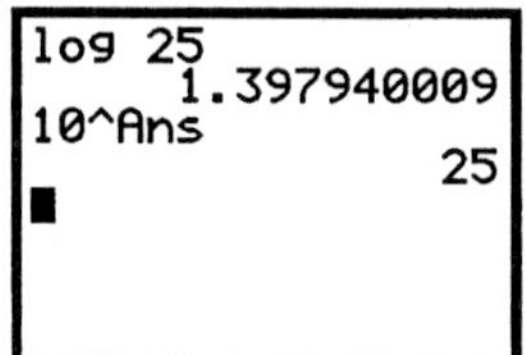

Figure 7. 12

Note: $\log 25 = \log_{10} 25$. This is the common logarithm. The base 10 is understood and conventionally not written.

Check your work:
$10^{1.397940009} = 25$

Press 10 $\boxed{\wedge}$ $\boxed{\text{2nd}}$ $\boxed{\text{ANS}}$ $\boxed{\text{ENTER}}$

Example 5

Sketch a graph of $y = \log x$ and use it to determine the domain and range of the function.

Press $\boxed{\text{Y=}}$ $\boxed{\text{CLEAR}}$ $\boxed{\text{LOG}}$ $\boxed{\text{X,T,}\Theta}$ (see Fig. 7.13).

Graph the function. Press $\boxed{\text{ZOOM}}$; select [4:Zdecimal].

Press $\boxed{\text{TRACE}}$. The graph in Figure 7.14 shows that the *log x* is undefined when $x = 0$ (see Fig. 7.14).

Use $\boxed{\triangleleft}$ to confirm that *log x* is undefined for $x \leq 0$ (see Fig. 7.15).

The domain of $y = \log x$ is the set of all x such that $x > 0$.

The range is evident by looking at the graph also. As x increases y increases, but what happens as x approaches zero? y seems to be headed in a negative direction. The range for $y = \log x$ is the set of all y such that y: $(-\infty$, $+\infty$), or y is any real number.

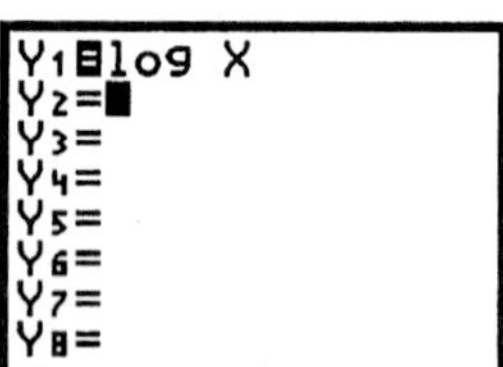

Figure 7. 13

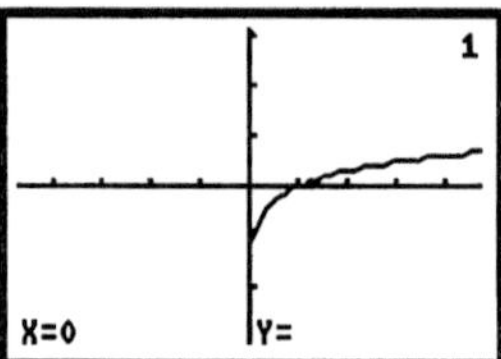

Figure 7. 14

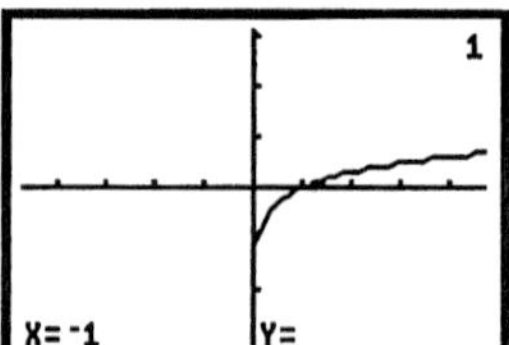

Figure 7. 15

Verify the range values by using $\boxed{\text{2nd}}$ $\boxed{\text{TblSet}}$ (see Fig. 7.16). Use $\boxed{\text{2nd}}$ $\boxed{\text{TABLE}}$ to see the values for $y = \log x$ for $0 < x < 1$, y is getting more negative (see Fig. 7.17).

7.4 Solving Equations Graphically

We saw in Chapter 6 that the point of intersection represented the solution to an equation. You can solve an equation graphically by locating the point of intersection.

Example 6

Solve the equation $10^x = 25$ graphically.

Enter the following into $\boxed{\text{Y=}}$:

$$Y1 = 10^x$$
$$Y2 = 25$$

See Figure 7.18.

Set the WINDOW so that both equations can be seen (see Fig. 7.19).

Press $\boxed{\text{GRAPH}}$ (see Figure 7.20).

The point of intersection represents the solution to the equation. Press $\boxed{\text{2nd}}$ $\boxed{\text{CALC}}$; select [5:intersect] (see Fig. 7.21). Follow the prompts by pressing $\boxed{\text{ENTER}}$.

The point of intersection occurs at approximately $x = 1.39794$ (see Fig. 7.22).

Note: The calculator remembers the intersection value for x. Immediately go to the Home Screen. Press $\boxed{\text{2nd}}$ $\boxed{\text{QUIT}}$; press $\boxed{\text{X,T,\Theta}}$ .See Figure 7.23 below.

Verify the solution by typing the expression 10^x (see Fig. 7.23 below).

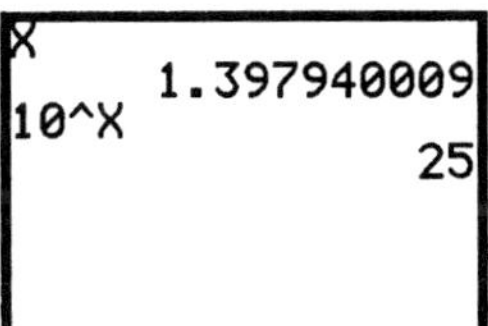

Figure 7. 23

Figure 7. 16

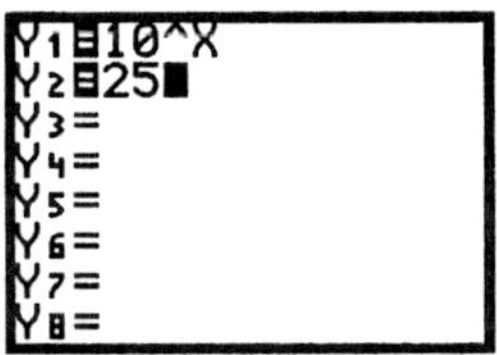

Figure 7. 17

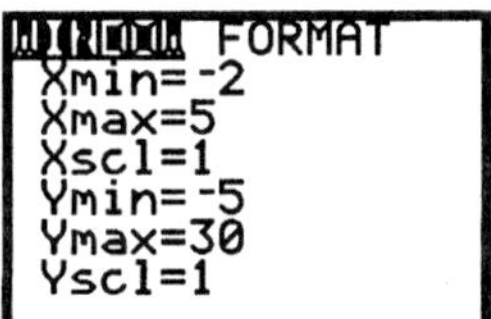

Figure 7. 18

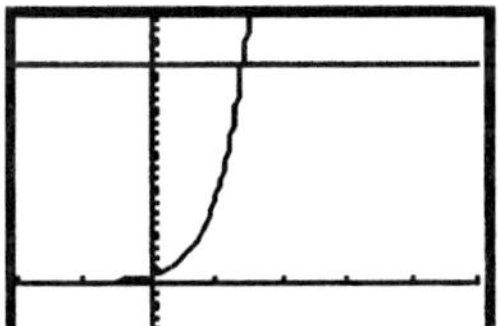

Figure 7. 19

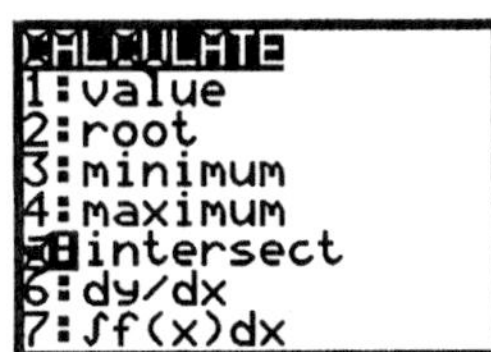

Figure 7. 20

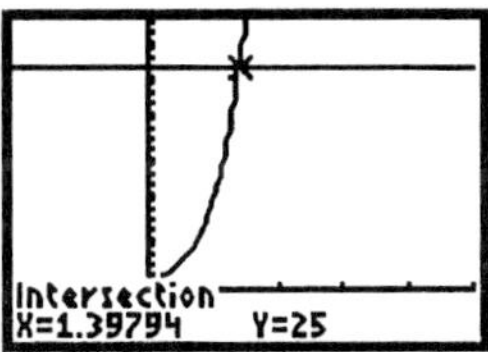

Figure 7. 21

Figure 7. 22

Chapter 8
Exponential Functions

8.1 Exponential Functions

Exponential functions are in the form
$y = b^x$ where $b > 0$, $b \neq 1$ and the power x
can be any real number.

Example 1

Graph the following in Y1, Y2, Y3, and Y4:

1. $f(x) = 2^x$ 3. $h(x) = 7^x$
2. $g(x) = 5^x$ 4. $k(x) = 10^x$

Type the functions into $\boxed{Y=}$ (see Fig. 8.1).

Set your $\boxed{\text{WINDOW}}$ as in Figure 8.2. Press

$\boxed{\text{GRAPH}}$ (see Fig. 8.3).

8.1.1 Exponential Growth

All of these graphs cross the y-axis at
$(0, 1)$, because $b^0 = 1$. These exponential
graphs have the same general shape. As x
increases y increases slowly at first and
then y increases very rapidly. The value of
y is increasing at an increasing rate. This
type of change is commonly called
exponential growth.

Set up a table of values (see Figs. 8.4 and
8.5). The table gives an idea of the growth
rate of $y1 = 2^x$. As x changes from 0 to 10, Y1
changes from 1 to about 1000. But as x
changes from 10 to 20, Y1 does not increase
to 2000, instead Y1 now changes from about
1000 to about 1,000,000! This is a magnitude
of 3 (3 powers of 10) or 10^3 times as large.

8.1.2 Exponential Decay

Examples 2

Graph the following Y1, Y2, Y3 and Y4:

1. $f(x) = 2^{-x}$ 3. $h(x) = 7^{-x}$
2. $g(x) = 5^{-x}$ 4. $k(x) = 10^{-x}$

Enter the functions as in Figure 8.6. These
exponential functions differ from those in
Example 1. A negative now appears in front
of the exponent x. Recall $b^{-x} = 1/b^x$. The
graphs are shown in Figure 8.7. The graphs
from Figure 8.3 are reversed. Now as x
increases y decreases very rapidly. This is
commonly called _exponential decay_.
Notice that as x gets larger y gets smaller
and gets very close to zero. All of these
exponential decay graphs have $(0,1)$ as the
y-intercept.

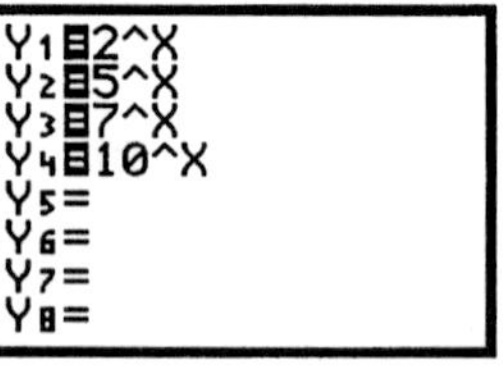

Figure 8. 1

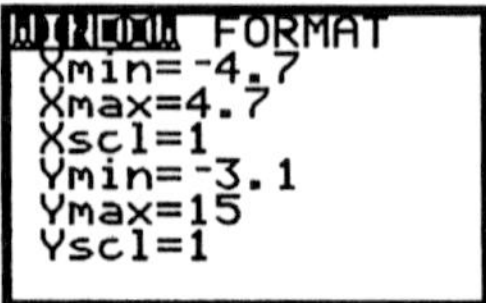

Figure 8. 2

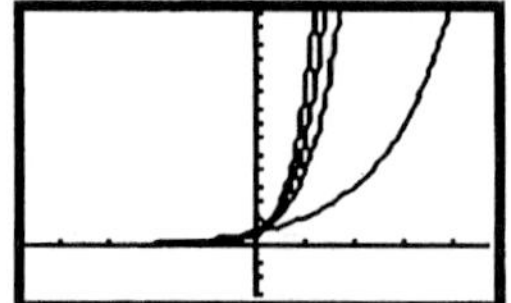

Figure 8. 3

Figure 8. 4

X	Y1	Y2
0	1	1
5	32	3125
10	1024	9.77E6
15	32768	3.1E10
20	1.05E6	9.5E13
25	3.36E7	3E17
30	1.07E9	9.3E20

X=20

Figure 8. 5

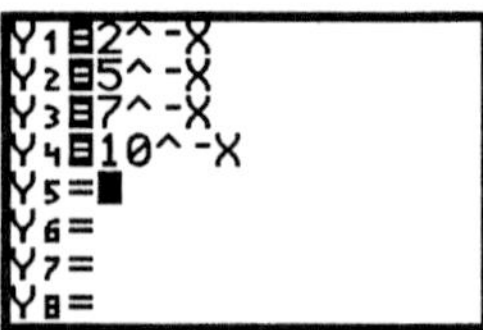

Figure 8. 6

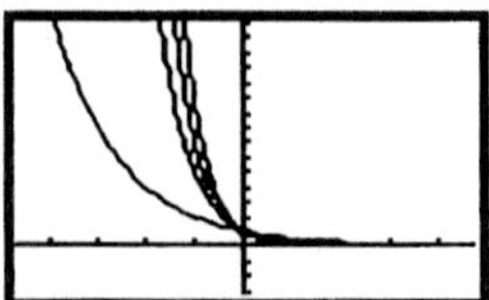

Figure 8. 7

Example 3

The population of a town with a current population of 1100 people is expected to grow at a rate of 5% per year. In how many years will the population double?

When the growth rate is a percent increase, this represents exponential growth. The function is : $P(x) = P_0 (1 + r)^x$. In this case the initial population is $P_0 = 1100$ and the rate of growth is $r = .05$.
Type the function $P(x) = 1100(1 + 0.05)^x$ in

$\boxed{Y=}$ (see Fig. 8.8).

Create a table of values. Press $\boxed{2nd}$ $\boxed{TblSet}$, begin with year 0 and set the increment (Δtbl=) to 1 year (see Figure 8.9). Press $\boxed{2nd}$

$\boxed{TABLE}$ (see Fig. 8.10).

Approximate the doubling time by using the table. Press $\boxed{\nabla}$ until y is approximately 2200, since $2P_0 = 2*1100 = 2200$. When $14 < x < 15$, y will be 2200 (see Fig. 8.11). This means that the population will double in a little more than 14 years.

Example 4

Solve the above problem graphically.
Let $P(x) = 2200$.
Find the solution to:
$$2200 = 1100(1 + 0.05)^x$$

Press $\boxed{Y=}$. Let Y2=2200 (see Fig. 8.12).
Adjust the Window based upon the table of values. x: [-10,20] and y: [-1000,2500].

Press $\boxed{GRAPH}$. Find the point of

intersection using the $\boxed{2nd}$ $\boxed{CALC}$

[5:intersect] menu (see Fig. 8.13).

Look at the graph of Y1 in Figure 8.13. The beginning population of 1100 is the y-intercept of the graph. The point of intersection represents the number of years it will take for the population to double, or y, to become 2200. This will occur when x is about 14.2 years.

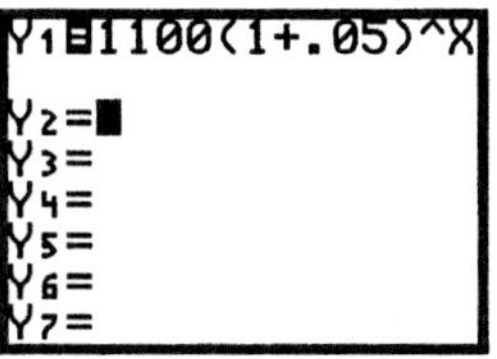

Figure 8. 8

Figure 8. 9

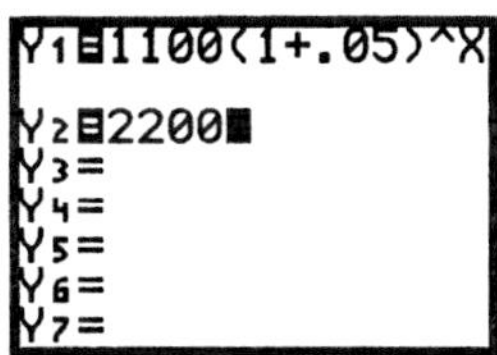

Figure 8. 10

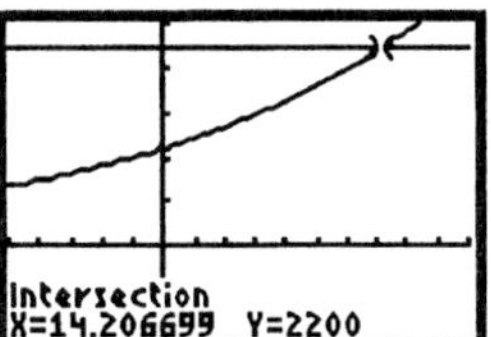

Figure 8. 11

Figure 8. 12

Figure 8. 13

Example 5
Graph the number of E. coli bacteria using a semi-log scale, if the bacteria grow according to the table below , where *t* is measured in 20 minute time periods.

time periods	number of E. coli
0	200
1	400
2	800
3	1,600
4	3,200
5	6,400
6	12,800
7	25,600
8	51,200
9	102,400
10	204,800

Table 8. 1

1. Use the statistics menu to graph the points from Table 8.1. Press $\boxed{\text{STAT}}$; select [1:Edit]; enter the time periods in L1 and enter the number of bacteria in L2 (see Fig. 8.14).

2. Find the log of list 2 and store to list 3. Move $\boxed{\Delta}$ to L3 label. Type $\boxed{\text{LOG}}$ $\boxed{\text{2nd}}$ $\boxed{\text{L2}}$ $\boxed{\text{ENTER}}$ (see Figs. 8.15 and 8.16).

3. Set up the plot. Press $\boxed{\text{2nd}}$ $\boxed{\text{STATPLOT}}$ $\boxed{\text{ENTER}}$. Set up the plot as in Figure 8.17.

4. $\boxed{\text{CLEAR}}$ $\boxed{\text{Y=}}$. Press $\boxed{\text{ZOOM}}$ select [9:ZoomStat] (see Fig. 8.18).

From the data it appears that our E. coli bacteria are growing exponentially because the number is doubling every 20 minutes. Since the number of bacteria is doubling (adding to itself), the growth rate of 100% represents a constant growth factor of 2. Therefore the plot of *x* against the *log y* forms a straight line. So $P = 200(1 + 1)^x$. (See Chapter 12 for more detail).

Note: Exponential data will appear linear when graphed on a semi-log scale

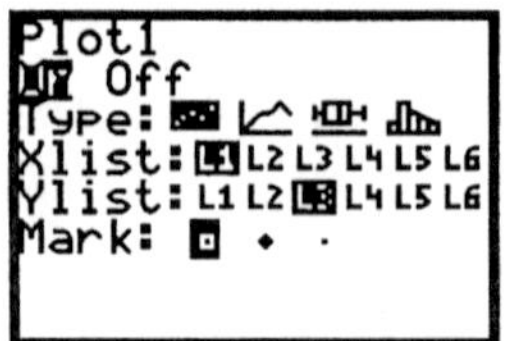

Figure 8. 14

Figure 8. 15

Figure 8. 16

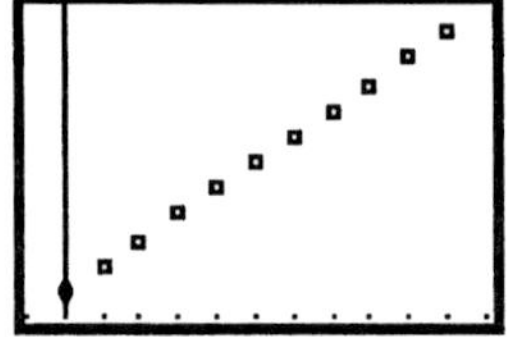

Figure 8. 17

Figure 8. 18

Example 6
Find the Linear regression equation for Example 5 , then give the algebraic explanation.
(You may have to refer to Chapter 12 for the property of logarithms to understand this example.)

Find the linear regression equation using the data in L1 and L3. From the Home Screen press $\boxed{\text{STAT}}$ <CALC>, select [5:LinReg(ax+b)] (see Fig. 8.19).
Press $\boxed{\text{2nd}}$ $\boxed{\text{L1}}$ $\boxed{,}$ $\boxed{\text{2nd}}$ $\boxed{\text{L3}}$ $\boxed{\text{ENTER}}$ (see Figs. 8.20 and 8.21). The equation is:
$$y = .301x + 2.301$$

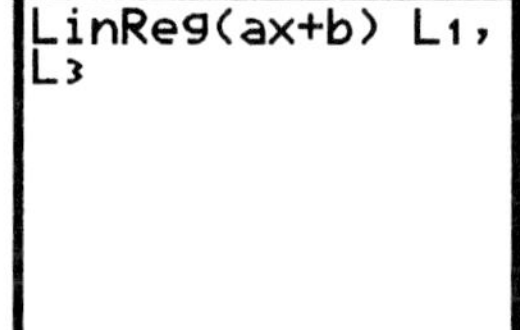

Figure 8. 19

The algebraic solution will tie the linear regression equation to the exponential function. Use properties of logarithms to simplify:
$$P = 200(2)^x$$
$$\log P = \log (200(2)^x)$$
$$\log P = \log 2^x + \log 200$$
$$\log P = (\log 2)x + \log 200$$

The final equation is in the linear form
$$y = mx + b.$$
$$m = \log 2 = .3010299957...$$
$$b = \log 200 = 2.30129996...$$
These are the same values as a and b in the regression equation (see Fig. 8.22).
Therefore x is the independent value and the $\log P = y$, the dependent value.

We find the equivalent antilog values are:
$10^{\log 2} = 2$ and $10^{\log 200} = 200$ (see Fig. 8.23).

Therefore the slope of the linear regression equation allows you to find the base of the exponential function and the y-intercept allows you to find the initial value of the exponential function.

Figure 8. 20

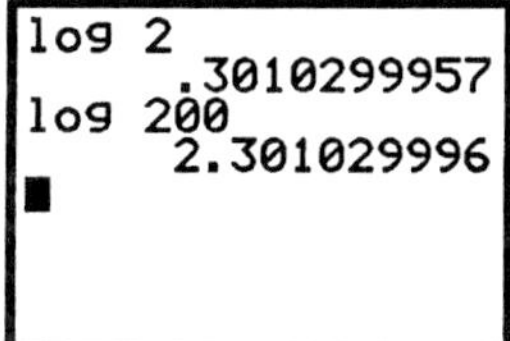

Figure 8. 21

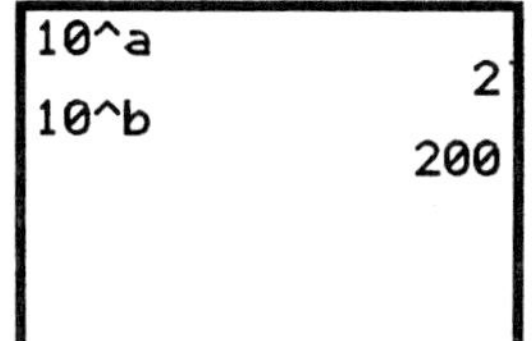

Figure 8. 22

Figure 8. 23

Chapter 9
Power Functions

9.1 Power Functions with Positive Integral Powers

A power Function has the form
$$y = kx^p$$
where k and p are constants. The simplest power function is $y = x$, where $p = 1$. Its graph is linear. However not all other power functions look linear.

9.1.1 Visualizing Odd Power Functions
Example 1
Graph the following and generalize the shape of the graph.

$$Y1 = x^3$$
$$Y2 = x^5$$
$$Y3 = x^7$$

Press $\boxed{Y=}$ $\boxed{CLEAR}$ to erase all old functions.

Enter the above functions (see Figure 9.1). Set your WINDOW as in Figure 9.2. Press $\boxed{GRAPH}$ (see Fig. 9.3).

The graphs go through the origin (0, 0), have both positive and negative y values and have a "lazy S " shape. As you move left to right the functions are always increasing.

A look at the table of values, in Figure 9.4, shows that for the odd power functions as x increases to $+\infty$, y increases pretty fast to $+\infty$. As x decreases to $-\infty$ y decreases to $-\infty$.

Press $\boxed{ZOOM}$; select [3:Zoom Out] (see Fig. 9.5). The calculator gives you a chance to reposition your cursor; then press $\boxed{ENTER}$ (see Fig. 9.6). This shows that all the graphs have a similar shape or "global behavior."

9.1.2 Visualizing Even Power Functions
Example 2:
Graph the following and generalize the shape of the graph.

$$Y1 = x^2$$
$$Y2 = x^4$$
$$Y3 = x^6$$

Figure 9. 1

Figure 9. 2

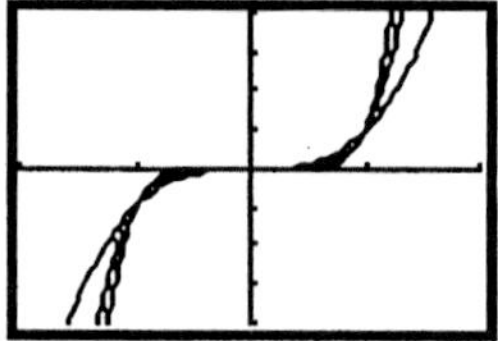

Figure 9. 3

Figure 9. 4

Figure 9. 5

Figure 9. 6

Enter the functions into $\boxed{Y=}$. Reset the WINDOW as in Figure 9.2 or press $\boxed{ZOOM}$ <Memory>; select [1:ZPrevious]. Press $\boxed{GRAPH}$ (see Figure 9.7).

The graphs go through the origin $(0, 0)$, have only positive y values and have a "U" shape. As you move from negative values to positive values of x the functions have large positive values then decrease to zero then increase again. A look at the table of values, in Figure 9.8, shows that for the even power functions, as x increases to $+\infty$, y increases pretty fast to $+\infty$. As x decreases to $-\infty$., y increases to $+\infty$.

9.1.3   Polynomial Functions

When positive integer power functions are added together you get a polynomial function:

$$y = a_n x^n + a_{n-1} x^{n-1} + ... + a_1 x^a + a_0$$

where a_n is a constant coefficient $(a_n \neq 0)$, and n is a positive integer power.

Example 3

The deer population in a national forest was monitored over a 25 year period. The data collected can be modeled by the fifth-degree polynomial:

$$D(x) = - 0.125x^5 + 3.125x^4 + 4000.$$

Graph the function and interpret the graph.

Let Y1= $-.125x^5 + 3.125x^4 + 4000$
(see Fig. 9.9).

Note: Use the table of values to find the appropriate WINDOW.

Press $\boxed{2nd}$ $\boxed{TblSet}$. Let TblMin = 0 and Δtbl = 1. Press $\boxed{2nd}$ $\boxed{TABLE}$ (see Figs. 9.10 and 9.11). It looks like the maximum population occurs around 20 years.
Set the WINDOW as in Figure 9.12.

Press $\boxed{GRAPH}$, then $\boxed{TRACE}$ to explore the graph (see Fig. 9.13).

Interpretation:
It took 20 years for the deer population to go from 4000 to a maximum population of about 104,000. Over the next five year period the population decreased back to 4000. The sharp decrease was probably a result of a lack of food supply caused by over population and/or disease.

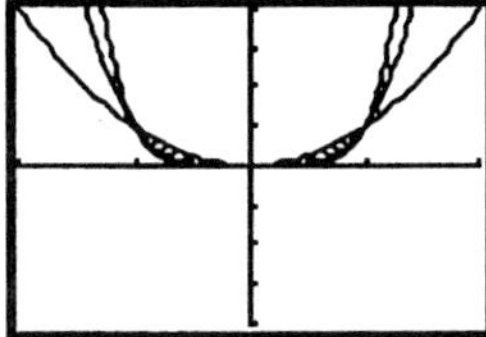

Figure 9. 7

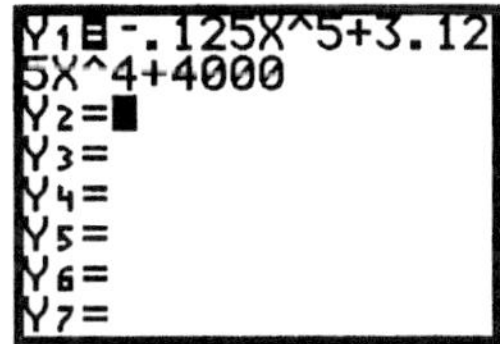

Figure 9. 8

Figure 9. 9

Figure 9. 10

Figure 9. 11

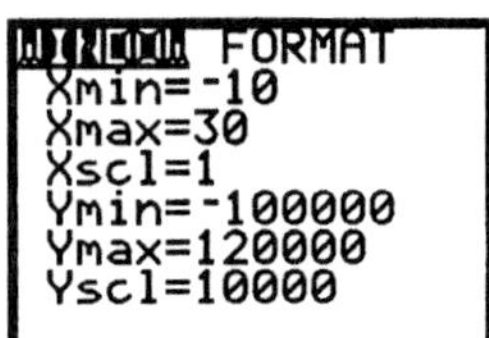

Figure 9. 12

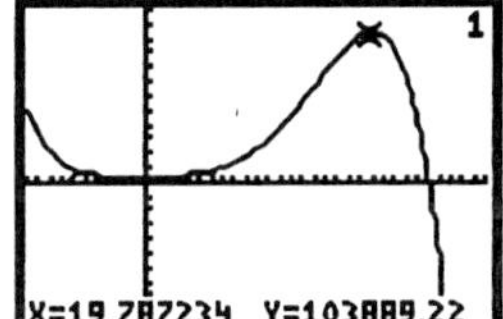

Figure 9. 13

9.1.4 Power Regression Equations.

Example 4

Below are some data reported on AIDS in women. Find a power regression equation that models the data.

Year	AIDS Cases
1	18
2	30
3	36
4	92
5	198
6	360
7	631
8	1016
9	1430

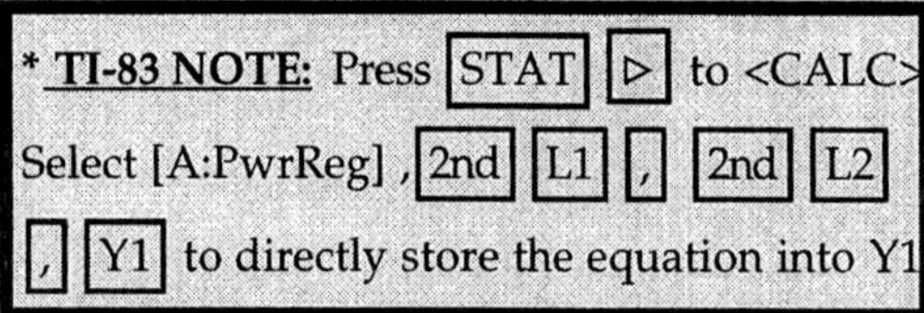

Note: Clear functions from $\boxed{Y=}$ and turn OFF all plots.

1. Enter the data in L1 and L2. Press $\boxed{STAT}$ [1:Edit] (see Fig. 9.14). Press $\boxed{2nd}$ $\boxed{STATPLOT}$ $\boxed{1}$ and set up the plot as in Figure 9.15. Press $\boxed{ZOOM}$; select [9:ZoomStat] (see Figs. 9.16 and 9.17).

2. Find the regression equation. Press $\boxed{STAT}$ $\boxed{\triangleright}$ to <CALC> ; select [B:PwrReg] $\boxed{2nd}$ $\boxed{L1}$ $\boxed{,}$ $\boxed{2nd}$ $\boxed{L2}$ $\boxed{ENTER}$ (see Figs 9.18 and 9.19).

3. Put the equation into Y1. To paste: *Press $\boxed{Y=}$ $\boxed{VARS}$; select[5:Statistics] $\boxed{\triangleright}$ $\boxed{\triangleright}$ to <EQ> ; select [:RegEQ]. Press $\boxed{GRAPH}$ (see Fig. 9.20).

*** TI-83 NOTE:** Press $\boxed{STAT}$ $\boxed{\triangleright}$ to <CALC> . Select [A:PwrReg] , $\boxed{2nd}$ $\boxed{L1}$ $\boxed{,}$ $\boxed{2nd}$ $\boxed{L2}$ $\boxed{,}$ $\boxed{Y1}$ to directly store the equation into Y1.

The power function fit is pretty good in the beginning, but not so good at the end.

Example 5

Find other polynomial regression equations (quadratic, cubic, or quartic) and an exponential equation .

Repeat steps 1 - 3 above choosing different regression equations. Then use these models to predict future AIDS values. Graphing screens will not be shown.

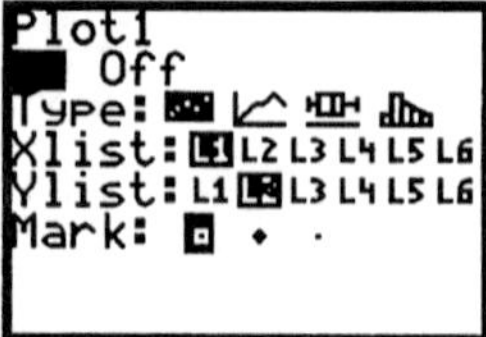

Figure 9. 14

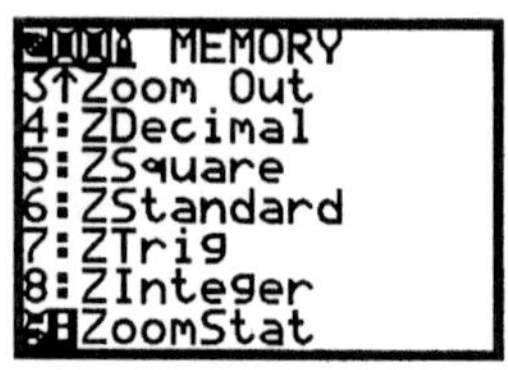

Figure 9. 15

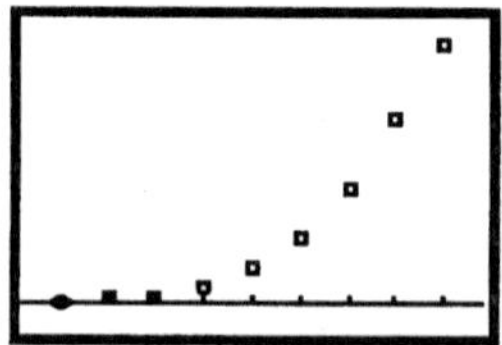

Figure 9. 16

Figure 9. 17

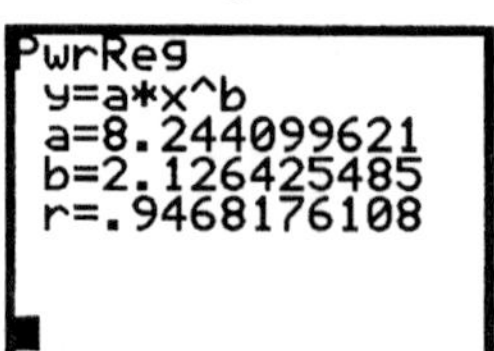

Figure 9. 18

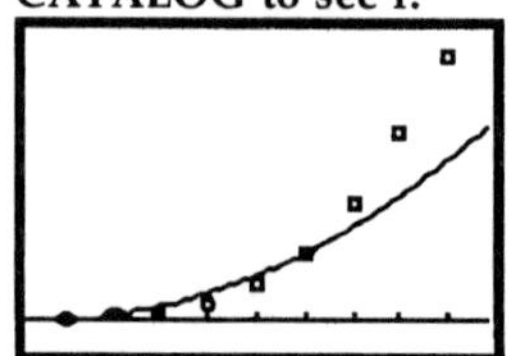

Figure 9. 19

TI-83. Turn DiagnosticsOn under CATALOG to see r.

Figure 9. 20

Example 6

Compare the graphs of negative integer power exponents and generalize about odd and even negative integer powers.

First Graph:

$$Y1 = x^{-1}$$

See Figure 9.21.

Press $\boxed{\text{ZOOM}}$ select [4:Decimal] (see Fig. 9.22). This graph is in two pieces, since $x^{-1} = 1/x$, so $x \neq 0$, and it is symmetric to the origin.

Graph other negative odd integer powers:

$$Y2 = x^{-3}$$
$$Y3 = x^{-5}$$

Enter the functions into Y1 and Y2. See Figure 9.23

Press $\boxed{\text{GRAPH}}$ (see Fig. 9.24)

To the right of $x = 0$, the function is at a large negative value and decreases rapidly to zero. As $x \to +\infty$, $y \to 0$.

To the left of $x = 0$, the function is near zero and decrease rapidly to large negative values. As $x \to -\infty$, $y \to 0$.

Graph even negative interger powers:

$$Y1 = x^{-2}$$

See Figure 9.25. Press $\boxed{\text{GRAPH}}$ (see Fig. 9.26). This is a reasonable graph because $x^{-2} = 1/x^2$, so $x \neq 0$ and all y values are positive. This graph is symmetric to the y -axis.

Graph other even negative integer powers:

$$Y2 = x^{-4}$$
$$Y3 = x^{-6}$$

Press $\boxed{\text{Y=}}$, enter the functions into Y2 and Y3. Press $\boxed{\text{GRAPH}}$ (see Fig. 9.27).

Once again the functions have the same general shape as x^{-2} with all positive values. On either side of $x = 0$, the function goes to a positive infinity. As one moves left to right the curve begins near zero and increases to $+\infty$. At $x = 0$,, the function is undefined. To the right of zero the function is at $+\infty$ and decreases rapidly to zero.

Troubleshooting: For x^{n} where n is a decimal value (non-integer) the domain of the function is limited to $x>0$. You will only see one-half of the power function.

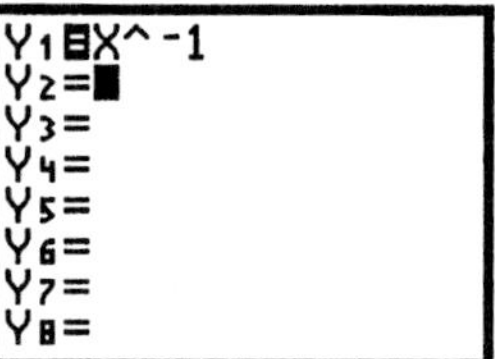

Figure 9. 21

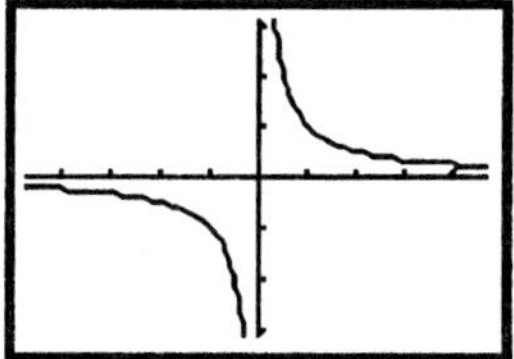

Figure 9. 22

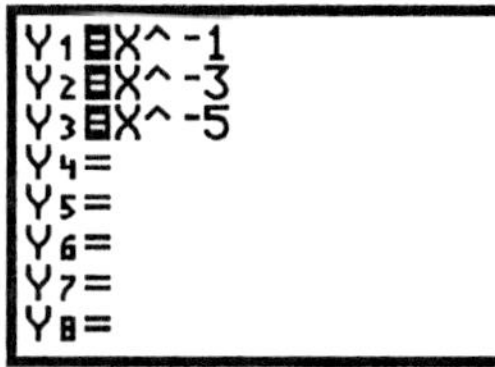

Figure 9. 23

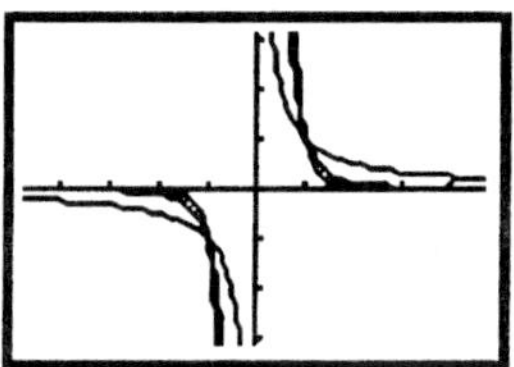

Figure 9. 24

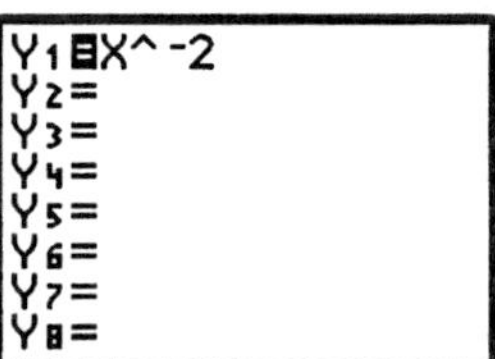

Figure 9. 25

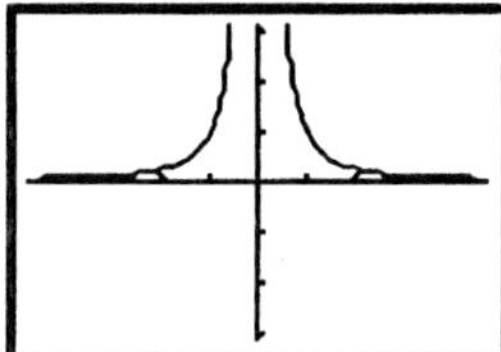

Figure 9. 26

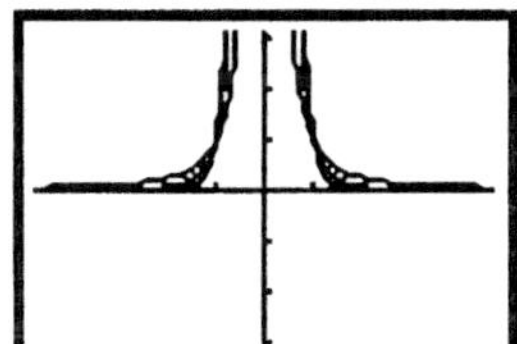

Figure 9. 27

59

Chapter 10
Free Fall Function Data

10.1   Free Fall Data

Can you think about how the height of an object varies over time when the object is dropped from a location like a window or bridge or even from your hand? If performing an experiment like dropping a ball in a classroom, the CBL™ (Calculator Based Laboratory) can be used. Instructions for using the CBL™ will be shared with you by your instructor or you can contact the WEB site: http://www.ti.com

Note: Be sure to review Chapter 5.1.1 Receiving Data.

The text comes with a free fall data program that can be transferred from calculator to calculator. Once the data is in your calculator either by experiment or from the program, do the following.
1.   Plot the data.
2.   Determine a quadratic regression equation for the data.

Figures 10.1 to 10.6 represent the calculator steps you need to perform.

Note: Remember to execute the program before you set up your plot. Press PRGM ; select FREEFALL ENTER . The program stores data to L1 - L2 (see Fig. 10.2)

You can also type the FREEFALL program (see below) into your calculator. Press

PRGM <NEW> ENTER ; type the program

name; ENTER ; and then type each line of

the program. Refer to your manual for more specifics. (See also the subsequent table)

```
PROGRAM:FREEFALL
:ClrList L₁,L₂,L₃,L₄,L₅,L₆
:{0.0000,0.0167,0.0333,0.0500,0.0
667,0.0833,0.1000,0.1167,0.1333,0
.1500,0.1667,0.1833,0.2000,0.2167
,0.2333,0.2500,0.2667,0.2833,0.30
00,0.3167,0.3333}→L₁
:{0.00,1.72,3.75,6.10,8.67,11.58,
14.71,18.10,21.77,25.71,29.90,34.
45,39.22,44.22,49.58,55.15,60.99,
67.11,73.48,80.10,87.05}→L₂
:{87.05,85.33,83.30,80.95,78.38,7
5.47,72.34,68.95,65.28,61.34,57.1
5,52.60,47.83,42.83,37.47,31.90,2
6.06,19.94,13.57,6.95,0.00}→L₃
```

Figure 10. 1

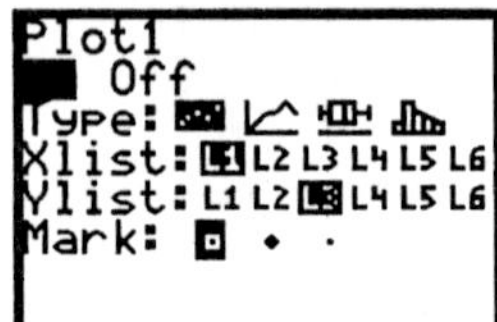

Figure 10. 2

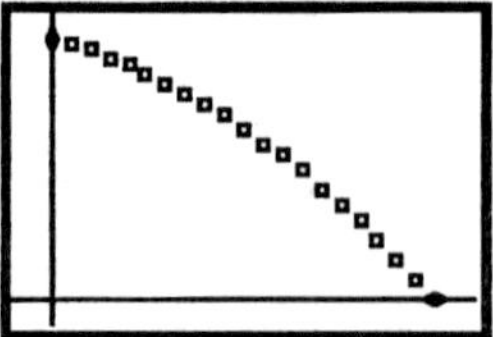

Figure 10. 3

Figure 10. 4

Figure 10. 5

Figure 10. 6

This data is also on the CD-ROM in the back of your text.

Distance fallen over time by an object in free fall			
Time	Distance fallen	Distance from Ground	
(sec)	(cm)	(cm)	
0.0000	0.00	87.05	
0.0167	1.72	85.33	
0.0333	3.75	83.30	
0.0500	6.10	80.95	
0.0667	8.67	78.38	
0.0833	11.58	75.47	
0.1000	14.71	72.34	
0.1167	18.10	68.95	
0.1333	21.77	65.28	
0.1500	25.71	61.34	
0.1667	29.90	57.15	
0.1833	34.45	52.60	
0.2000	39.22	47.83	
0.2167	44.22	42.83	
0.2333	49.58	37.47	
0.2500	55.15	31.90	
0.2667	60.99	26.06	
0.2833	67.11	19.94	
0.3000	73.48	13.57	
0.3167	80.10	6.95	
0.3333	87.05	0.00	

Chapter 11
Quadratic and Polynomial Functions

11.1 Quadratic Functions

Functions of the form $y = ax^2 + bx + c$ are called quadratic or second degree equations.

> **Note**: The highest power of x is two. All quadratic equations have the shape of a parabola.

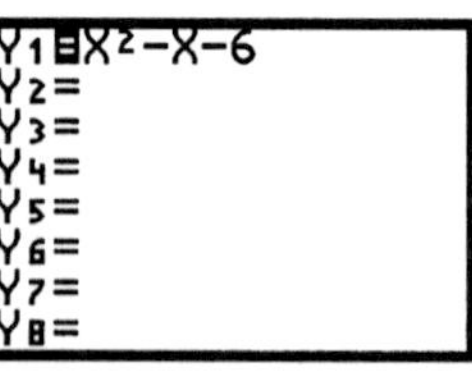

Figure 11. 1

Example 1

Graph $y = x^2 - x - 6$ and identify a, b, and c.

For this quadratic equation:

$$a = 1, b = -1 \text{ and } c = -6$$

To graph press $\boxed{Y=}$ $\boxed{CLEAR}$. Into Y1 enter $x^2 - x - 6$. Press $\boxed{ZOOM}$ $\boxed{6}$, since a=1 the graph opens up (see Fig. 11.1).

11.1.1 Finding y-intercept.

Press $\boxed{TRACE}$ to find the y-intercept; if $x = 0$ then $y = -6 = c$ (see Fig. 11.2).
The y-intercept is (0, -6) = (0, c).

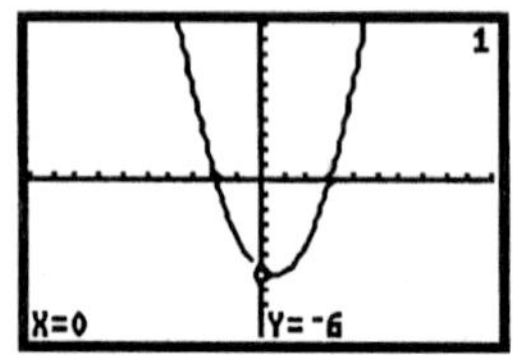

Figure 11. 2

11.1.2 Finding x-intercepts / Finding the Roots/ Finding the zeros.

Use $\boxed{TRACE}$ and arrows to estimate where the graph crosses the x-axis. There are two *x-intercepts* (where $y = 0$), also known as *roots* or *zeros*. The left intercept is around $x = -2$ and the right intercept is around $x = 3$. (see Figs. 11.3 and 11.4).
Use the calculator to find the left x-intercept (see Fig. 11.5) by following the steps below (see Figs. 11.6 -11.11 for TI-82, TI-83 differences).

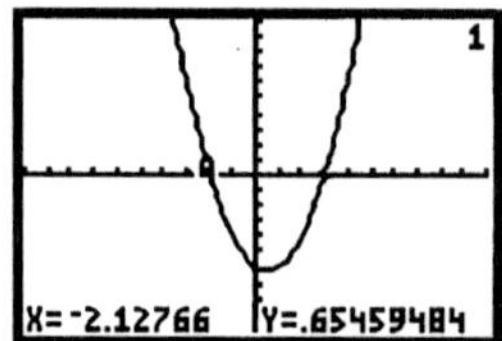

Figure 11. 3

Press $\boxed{2nd}$ $\boxed{CALC}$ $\boxed{2}$. Follow these steps:

1. Position the cursor **to the left** of the x - intercept when prompted. Press $\boxed{ENTER}$

2. Reposition the cursor **to the right** of the x -intercept when prompted. Press $\boxed{ENTER}$.

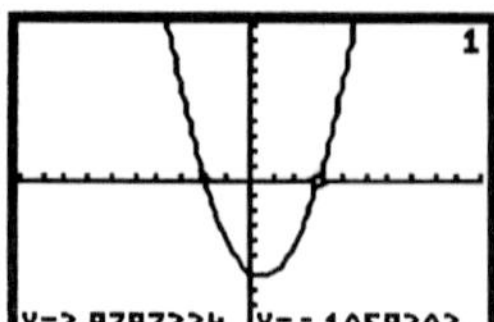

Figure 11. 4

3. When prompted GUESS? position the cursor near the x -intercept. Press $\boxed{ENTER}$. The root or x -intercept value is displayed (see Fig. 10.5).

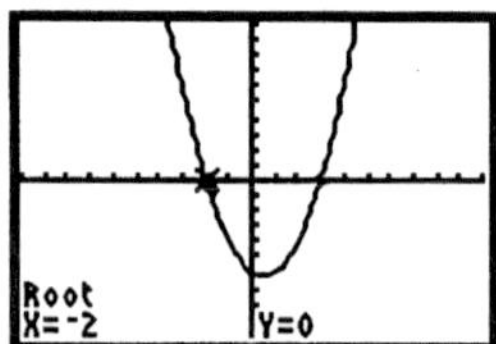

Figure 11. 5

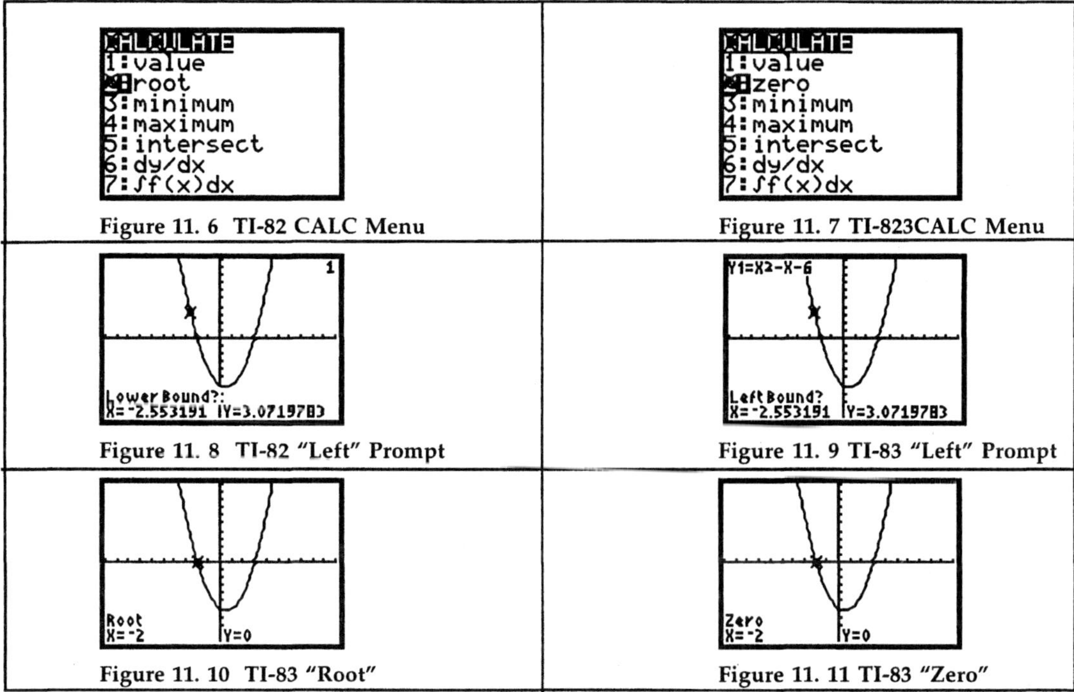

Figure 11. 6 TI-82 CALC Menu

Figure 11. 7 TI-823CALC Menu

Figure 11. 8 TI-82 "Left" Prompt

Figure 11. 9 TI-83 "Left" Prompt

Figure 11. 10 TI-83 "Root"

Figure 11. 11 TI-83 "Zero"

Find the right x - intercept:
Repeating the process we find that the other x - intercept is $(3, 0)$(see Fig. 11.12). There are two roots, at $x = -2$ and $x = 3$.

11.1.3   Check Roots Algebraically:
$$y = x^2 - x - 6$$
if $x = -2$, $y = (-2)^2 - (-2) - 6 = 0$
if $x = 3$, $y = 3^2 - 3 - 6 = 0$ (see Fig. 11.13).

11.1.4   Finding the Vertex of a Quadratic Function

We see from the graph in Figure 11.12 that the function $y = x^2 - x - 6$ opens upward. This will be true when $a > 0$. As we move left to right the *minimum* point on the graph is called the *vertex*. It is somewhere between $x = 0$ and $x = 2$.

Example 2

Find the vertex for $y = -2x^2 -12x -13$.

Notice that $a < 0$ and the graph opens downward (see Figs. 11.14 and 11.15). Now the vertex is the *maximum* point on the graph. The vertex appears to be around the point $(-3, 5)$.

Use the calculator to find the vertex.

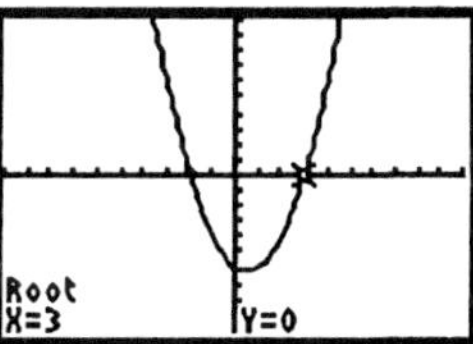

Figure 11. 12

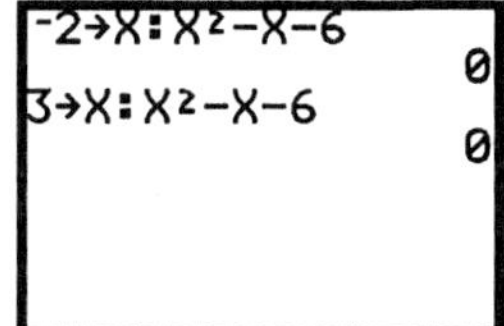

Figure 11. 13

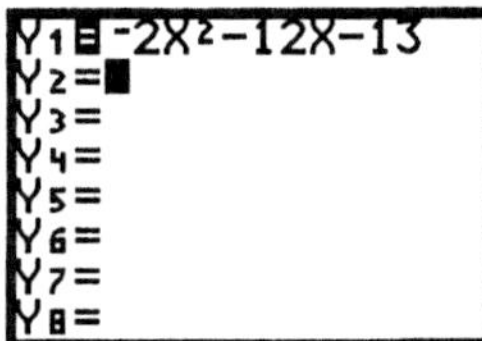

Figure 11. 14

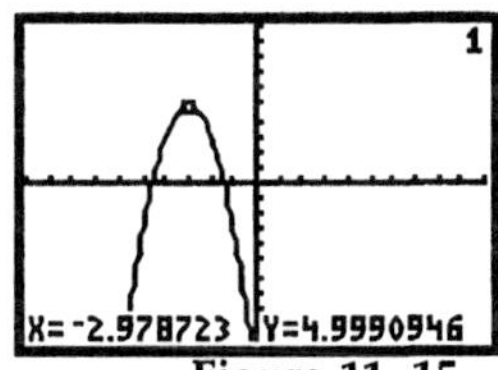

Figure 11. 15

Press $\boxed{2nd}$ $\boxed{CALC}$ $\boxed{4}$ for maximum (see Fig. 11.16). Then follow these steps*:
1. Position the cursor ***to the left*** of the maximum (lower bound) when prompted. Press $\boxed{ENTER}$ (see Fig. 11.17).
2. Reposition the cursor ***to the right*** of the maximum (upper bound) when prompted (see Fig. 11.18). Press $\boxed{ENTER}$

When prompted GUESS? position the cursor near the maximum. Press $\boxed{ENTER}$ (see Fig. 11.19).
The vertex is at the point (-3, 5) (see Fig. 11.20).

*__TI-83 Note:__ As with the x - intercept the TI-83 prompts for left and right bounds. You can also enter a number guess for the maximum. Only the TI-82 screens are shown to the right.

11.1.5 The Vertex Form of a Quadratic Function.

We write the vertex as the point (h, k).
Then the quadratic equation $y = ax^2 + x + c$
is transformed to :
$$y = a(x - h)^2 + k$$
So $y = -2x^2 -12x -13$ becomes
$$y = -2(x - {}^-3)^2 + 5$$
$$= -2(x + 3)^2 + 5$$

Verify algebraically:
$$y = -2(x + 3)^2 + 5$$
$$= -2(x^2 + 6x + 9) + 5$$
$$= -2x^2 -12x - 18 + 5$$
$$= -2x^2 -12x - 13$$

Verify Numerically:

Enter both equations into $\boxed{Y=}$ (see Fig. 11.21). Set up a table of values. Press $\boxed{2nd}$ $\boxed{TblSet}$ (see Fig. 11.22).

Press $\boxed{2nd}$ $\boxed{TABLE}$.

Figure 11.23 shows that for all values of x , Y1 and Y2 are equivalent.

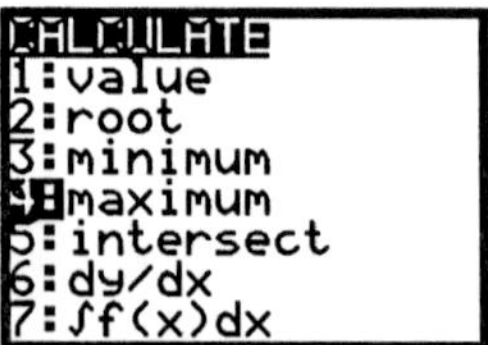
Figure 11. 16

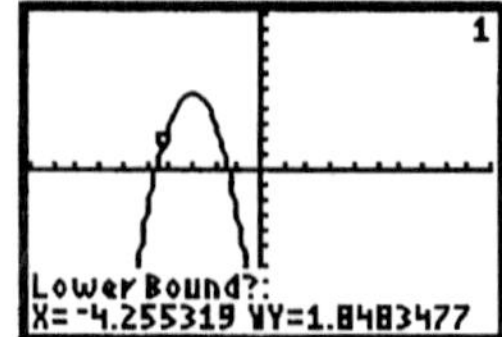
Figure 11. 17

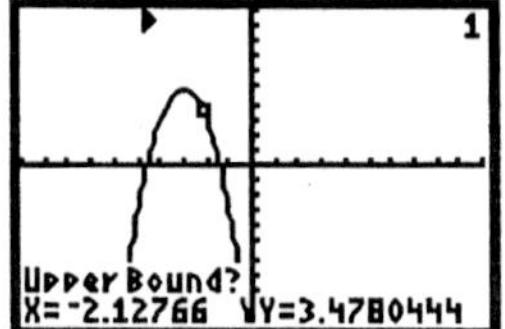
Figure 11. 18

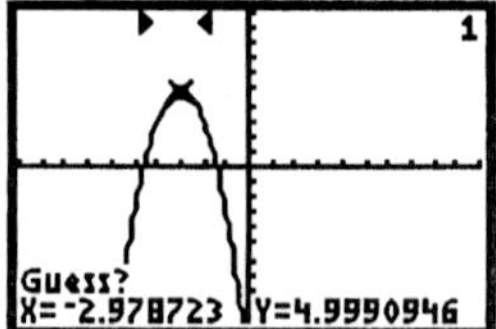
Figure 11. 19

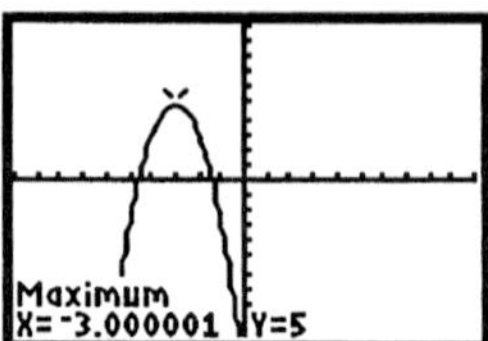
Figure 11. 20

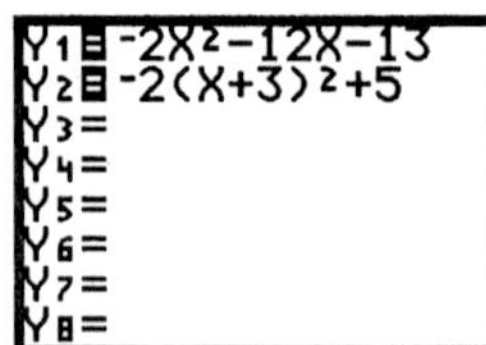
Figure 11. 21

Figure 11. 22

Therefore Y1 = Y2, or
$$-2x^2 - 12x - 13 = -2(x+3)^2 + 5 \text{ and}$$
$$ax^2 + bx + c = a(x-h)^2 + k$$

11.1.6 Finding an Appropriate Window.
Example 3
Graph $y = 3x^2 - 20x + 45$.

Into Y1 enter: $3x^2 - 20x + 45$.

CLEAR all other functions.

Press ZOOM 6 . We see nothing!

Use the table to get a feel for what
happens to y as x increases.

Press 2nd TABLE . Figure 11.24 shows that
when $x = -5$, $y = 220$ and when $x=0$, $y= 45$.
No wonder we couldn't see anything on a
[-10,10] by [-10,10] standard window.
From algebra we know that $a = 3 > 0$ so the
graph opens upward. Use down arrow to
find the minimum value of y (the vertex).
From the table (see Fig. 11.25) it looks like
the minimum occurs around $x = 3$.
Adjust the window so that you can see the
point (3, 12) as well as the point (-5, 220).

Press WINDOW ▽ ; let Ymax = 300 (see

Fig. 11.26).

Press GRAPH (see Fig. 11.27). It looks like

the graph is being cut off on the right side.
We need to see more values of x. Press

GRAPH . Change Xmax to 15.

Press GRAPH (see Fig. 11.28).

11.1.7 A Complete Graph

Try to select a window that displays a
complete graph. A complete graph shows
the whole shape of the graph, with all its
turning points and end behavior. Also
shown are the y -intercept and x -
intercept(s), if they exist. There are many
complete graphs.

Figure 11.28 shows a complete graph of
$$y = 3x^2 - 20x + 45$$

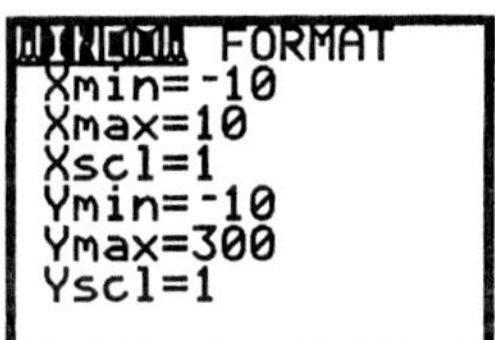

Figure 11. 27

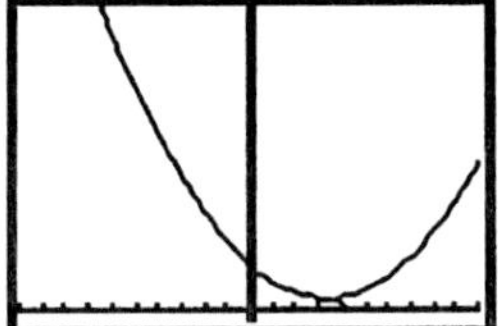

Figure 11. 28

11.2 Polynomial Functions

In Chapter 9 you were introduced to power functions. When positive integer power functions are added together you get a polynomial function:

$$y = a_n x^n + a_{n-1} x^{n-1} + \ldots + a_1 x^a + a_0$$

where a_n is a constant coefficient $(a_n \neq 0)$, and n is a positive integer power.

Example 4

The fall term enrollment of a university's freshman across the years is given below. Find a regression model that would predict the enrollment in 2000 if trends continue

year	freshman
1986	8662
1987	8986
1988	9416
1989	9780
1990	9936
1991	10050
1992	9977
1993	9846
1994	9570
1995	9582
1996	9610
1997	9693

Press STAT [1:Edit]. Enter the years into L1 and the number of freshman in L2 (see Fig. 11.29). Press ZOOM [9:ZoomStat] (see Fig. 11.30). The data appears to increase to 10,050 then decrease to 9,570 then increase again. A polynomial regression equation with at least two turning points is a cubic. Find the cubic regression model.

Press STAT <CALC>; select [:CubicReg] as in Figure 11.31. Type the list names. Press 2nd L1 , 2nd L2 ENTER .

Figure 11.32 shows all the coefficients of the thrid degree polynomial, or cubic function. Store the equation to Y1; press Y=

VARS select[5:Statistics] <EQ> ; select [:RegEQ] (see Fig. 11.33). Press GRAPH (see Fig. 11.34). To find the predicted number of freshman in 2000, store 2000 to x. The value of Y1 is 10457 freshman in the year 2000 (see Fig. 11.35).

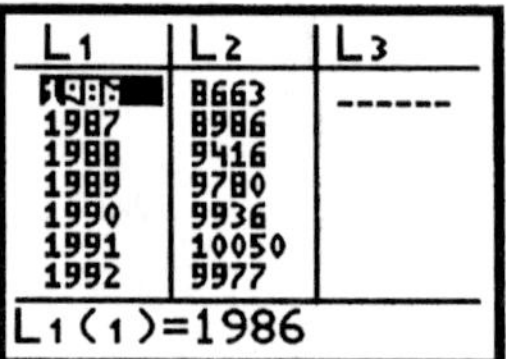

Figure 11. 29

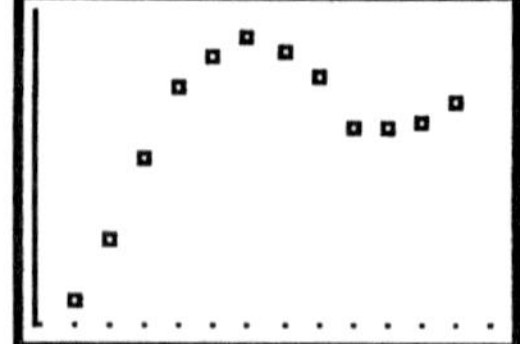

Figure 11. 30

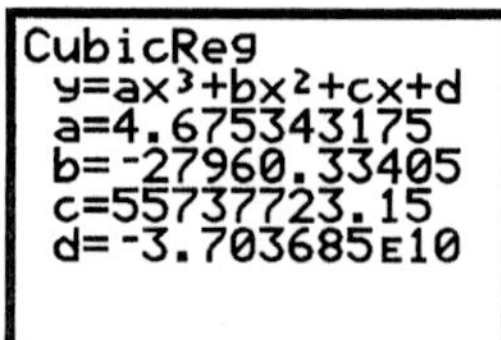

Figure 11. 31

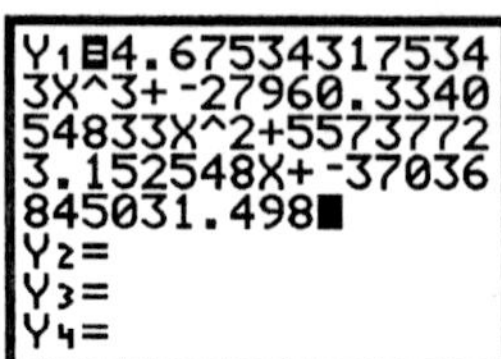

Figure 11. 32

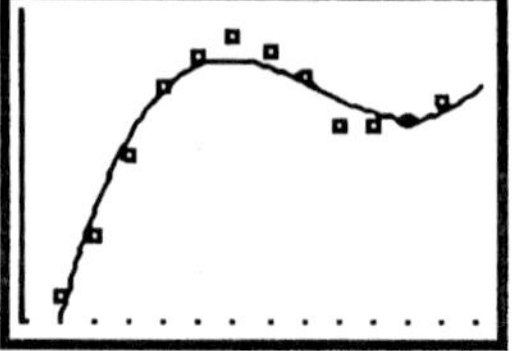

Figure 11. 33

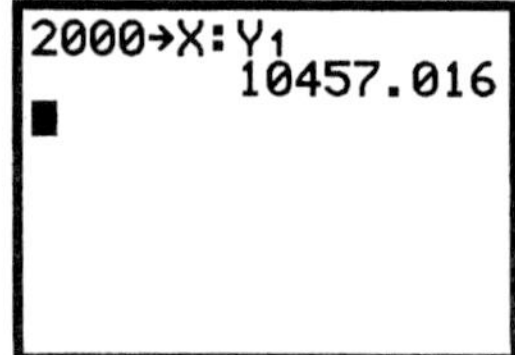

Figure 11. 34

Figure 11. 35

Chapter 12
Logarithmic Links:
Logarithmic, Exponential and Power Functions

12.1   The Definition of e
Example 1
Consider the exponential function

$$f(x) = \left(1 + \frac{1}{x}\right)^x$$

What happens to $f(x)$ as x gets very large?

Type the function into $\boxed{Y=}$ (see Fig. 12.1).

12.1.1   Use TABLE to Find values

Press $\boxed{2nd}$ $\boxed{TblSet}$. Put the independent

variable in ASK mode (see Fig. 12.2).

Press $\boxed{2nd}$ $\boxed{TABLE}$.

Type 100 $\boxed{ENTER}$ (see Fig. 12.3). The

calculator will only find the values of Y1
as you enter x values.
Continue to enter powers of 10 values as in
Figure 12.4.

Interpretation:
As $x \to +\infty$, $f(x) \to 2.718\ 281\ 8271...$ We
say that the function approaches a *limit*, or
gets close to a number value.

Example 2
**Compare the limit value of $f(x)$ to the
calculator value of e.**

Press $\boxed{2nd}$ $\boxed{QUIT}$ $\boxed{CLEAR}$. Store 10^9 to x.

Type Y1 and e^1 (see Fig. 12.5).

Note: e is located above $\boxed{LN}$ and you must
supply a power.

Notice that are nearly the same. The
value of e is an irrational number or

$$\lim_{x \to \infty} \left(1 + \frac{1}{x}\right)^x = e \approx 2.781828...$$

12.2   Draw the Inverse
Graphically speaking the inverse of a
function is a reflection about the line $y = x$.
The effect is that any point (a, b) becomes
(b, a) on the inverse graph.

Example 3
Graph $f(x) = e^x$ and its inverse , $f^{-1}(x)$.

Press $\boxed{Y=}$ enter e^x into Y1. Press $\boxed{ZOOM}$
[4:ZDecimal].

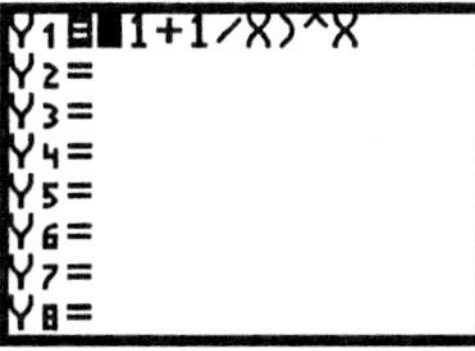

Figure 12. 1

Figure 12. 2

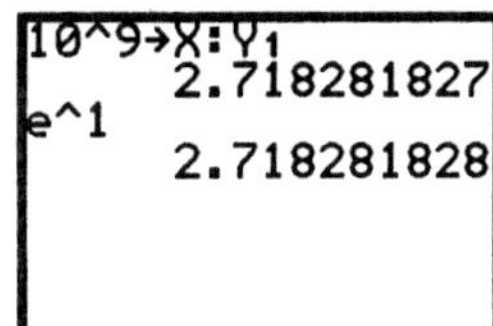

Figure 12. 3

X	Y1	
100	2.7048	
10000	2.7181	
100000	2.7183	
1E6	2.7183	
1E7	2.7183	
1E8	2.7183	
1E9		

Y1=2.7182818271

Figure 12. 4

10^9→X:Y1
         2.718281827
e^1
         2.718281828

Figure 12. 5

Use the : to type two commands.

To draw the inverse:

1. Press [2nd] [DRAW] ; select [8:DrawInv]
   (see Figs 12.6 and 12.7).

2. Type the function e^x. Press [ENTER].

   After pressing ENTER, the calculator
   automatically goes to the graph screen.
   (see Fig. 12.8).

> **Trouble Shooting:** Since you are in the draw
> mode you **can not** TRACE on the inverse
> function. To CLEAR the drawing press [2nd]
> [DRAW] select [1:ClrDraw].

12.1.3 Algebraically Finding the Inverse Function and Graphing $f^{-1}(x)$.

Example 4

Find $f^{-1}(x)$ for $y = e^x$ and graph $f^{-1}(x)$.

1. Use the definition of logarithm to "undo" the exponential function.

   if $\quad y = e^x$

   then $\quad \log_e y = x$

2. Reverse x and y to graph the function on the x, y coordinate plane:

   $$y = \log_e x = \ln x$$

   Using function notation:

   if $\quad f(x) = e^x$

   then $\quad f^{-1}(x) = \ln x.$

3. Graph $y = \ln x$.

Press [Y=] into Y2; type [LN] [X,T,Θ] (see Fig. 12.9).

Press [GRAPH] (see Fig. 12.10).

> **Note:** Now you can TRACE on the inverse
> graph. The point (1, 0) on graph Y1 is the
> point (0, 1) on the inverse graph, Y2.

12.2 Graphing Common and Natural Logarithms

The graphing calculator has two built-in logarithms:

1. For the common logarithm, $(\log_{10} x)$, use the [LOG] key.

2. For the natural logarithm, $(\log_e x)$, use the [LN] key.

Example 5

Compare the graphs of :

$$f(x) = \log x \quad \text{and} \quad g(x) = \ln x.$$

Enter the functions into Y1 and Y2 then
select the window with x:[0,4.7] and
y: [-4,4] (see Figures 12.11 and 12.12).

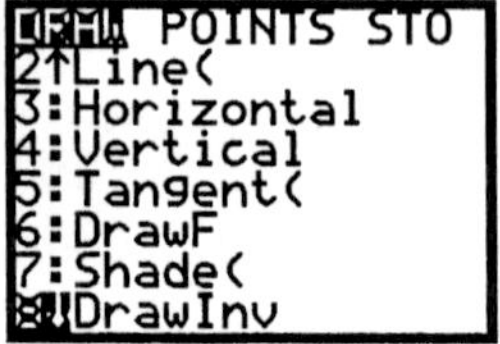

Figure 12. 6

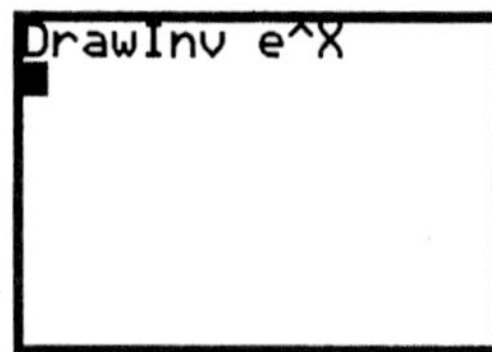

Figure 12. 7

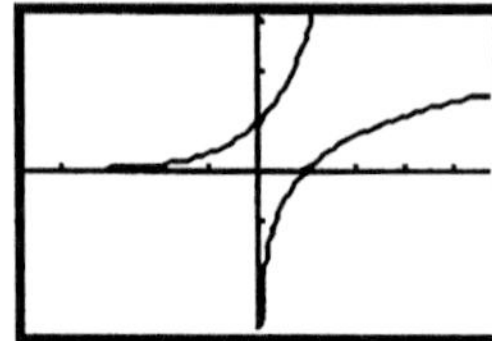

Figure 12. 8

Figure 12. 9

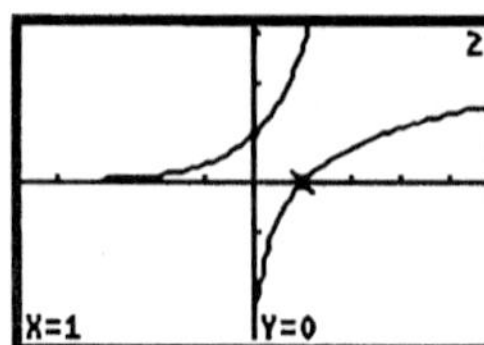

Figure 12. 10

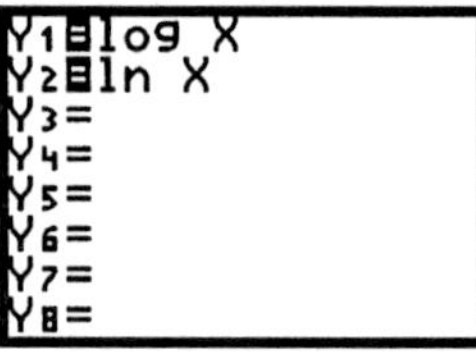

Figure 12. 11

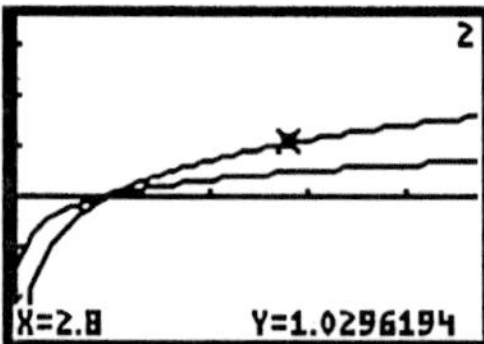

Figure 12. 12

Interpretation

Since the logarithm is the inverse of the an exponential function, the value of the logarithm represents the exponent of an exponential function. Thus, the larger the base of the logarithm the slower the exponent grows. The function *ln x* grows faster than *log x*. The graph of *ln x* is on top of *log x* for $x > 1$.

12.2.1 Graphs of Logarithms to Bases Other Than 10 and *e*.

Example 6

Find the inverse of f(x) = 3^x and graph it.

1. By definition $f^1(x) = \log_3 x$, which is not a built function on the calculator.
2. Use Algebra and properties of logarithms to find the inverse function.

$$
\begin{aligned}
y &= 3^x \\
x &= 3^y &&\text{exchange } x \text{ and } y \\
\log x &= \log 3^y &&\text{take the } log \text{ of both sides} \\
\log x &= y \log 3 &&\text{exponent property} \\
\log x / \log 3 &= y &&\text{divide by } log\ 3 \\
y &= \log x / \log 3
\end{aligned}
$$

> **Note**: The change of base formula can be used, $\log_b N = \log N / \log b = \ln N / \ln b$

3. Graph both 3^x and *log x/log 3* , then compare them.

    Press Y= enter the functions (see Fig. 12.13)

    Press ZOOM ; select [4:Zdecimal] (see Fig. 12.14). TRACE to $x = 0.7$

4. Check numerically:

    Y1=3^7 = 2.15766928

    Y2 = *log* 2.15766928 / *log* 3 = .7

    See Figure 12.15.

Verify the point on the inverse graph.

Press 2nd   CALC ; select [1:value].

Type in 2.15766928 ENTER   (see Fig. 12.16).

V takes you to the point on the inverse graph (see Figure 12.17).

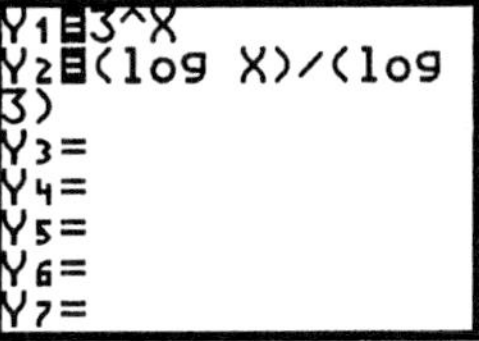

Figure 12. 13

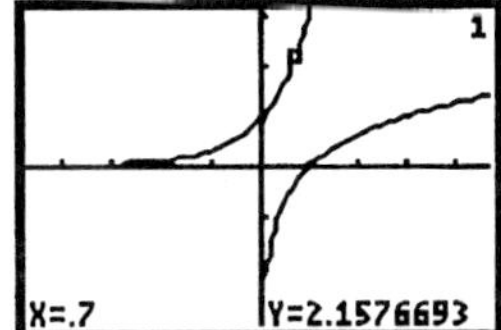

Figure 12. 14

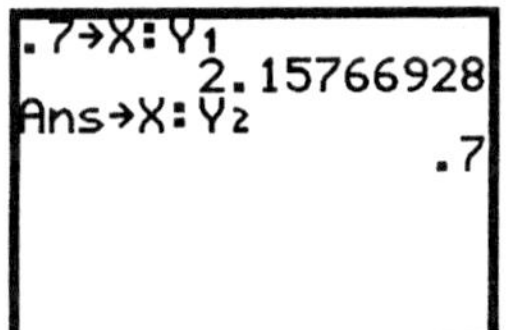

Figure 12. 15

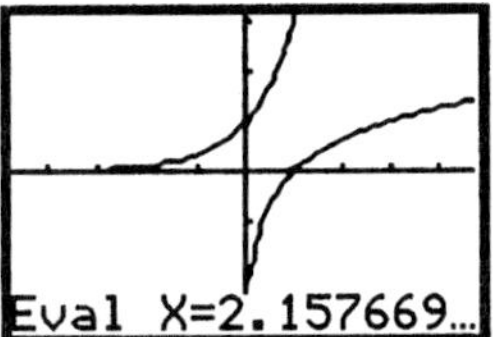

Figure 12. 16

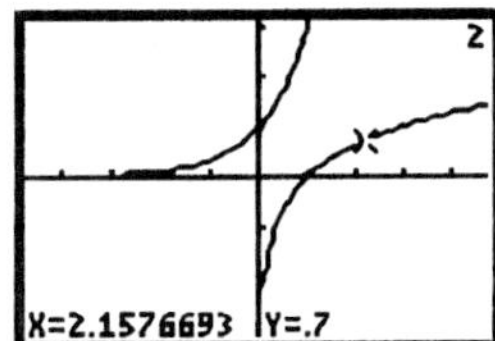

Figure 12. 17

12.3 Log -log Plots

When data is collected and graphed by hand, log-log paper is sometimes used. We will now explore how power functions behave on log-log plots.

Example 7

7.1. Graph the power function data and its formula:

L1	1	2	3	4	5	6	7
L2	1	8	27	64	125	216	343

1. Press STAT ; select [1:Edit]. Enter the data into L1 and L2 (see Fig. 12.18). Notice that L2 = L1^3. A function that would describe this relation is $y = x^3$.

2. Into Y= type x^3 (see Fig. 12.19).

3. Set up the plot (see Fig. 12.20). Press ZOOM ; select [9:ZoomStat].

7.2 Graph the data as a log-log plot.

1. Let log L1 = L3 and log L2 = L4 (see Fig. 12.22)

2. CLEAR Y=

3. Set up the plot with L3 and L4 (see Fig. 12.23)

4. Press ZOOM ; select [9:ZoomStat] (see Fig. 12.24).

Verify algebraically:

$$y = x^3$$

$$\log y = \log x^3$$

$$\log y = 3\log x$$

When you graph this equation, $log\ x$ is the independent variable (input value) and $log\ y$ is the dependent variable (output value). Thus we have a direct proportion with the 3 representing the slope of a line. Thus the slope will always be the power of the power function.

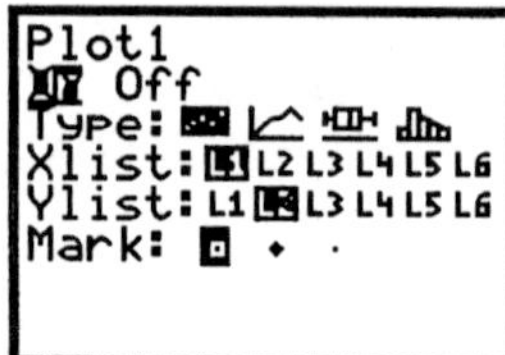

Figure 12. 18

Figure 12. 19

Figure 12. 20

Figure 12. 21

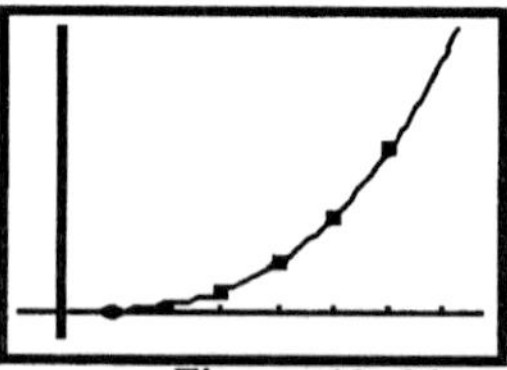

Figure 12. 22

Figure 12. 23

Figure 12. 24

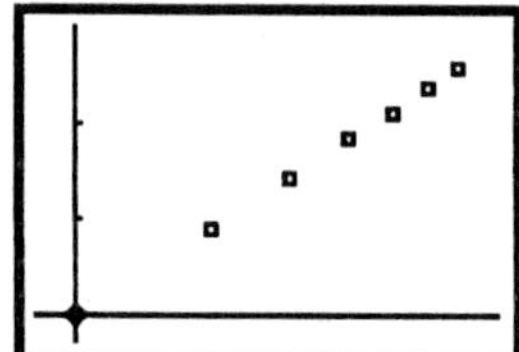